Marine Fishes and
Fisheries of New York

Marine Fishes and Fisheries of New York

HOWARD M. REISMAN

AND EMERSON C. HASBROUCK JR.

PHOTOGRAPHS BY CHRISTOPHER PAPARO

TECHNICAL EDITOR: RUSTY TYLER

Comstock Publishing Associates
an imprint of
Cornell University Press
Ithaca and London

Publication of this book was made possible by a generous grant from
Cornell Cooperative Extension of Suffolk County, Marine Program,
in support of fisheries educational outreach.

Underwater photography © Christopher Paparo

First published 2024 by Cornell University Press.

Design and composition by Julie Allred, BW&A Books, Inc.
Printed in China

Library of Congress Cataloging-in-Publication Data
Names: Reisman, Howard M., 1937– author. | Hasbrouck, Emerson C., Jr.,
    1951– author. | Paparo, Christopher, 1976– photographer.
Title: Marine fishes and fisheries of New York / Howard M. Reisman and
    Emerson C. Hasbrouck Jr., photographs by Christopher Paparo.
Description: Ithaca [New York] : Comstock Publishing Associates, an imprint
    of Cornell University Press, 2024. | Includes bibliographical references
    and index.
Identifiers: LCCN 2023045613 | ISBN 9781501774102 (paperback)
Subjects: LCSH: Marine fishes—New York (State) | Fisheries—New York
    (State) | Warmwater fishes—New York (State)
Classification: LCC QL628.N7 R45 2024 | DDC 597.177097474/1—
    dc23/eng/20240125
LC record available at https://lccn.loc.gov/2023045613

Contents

Acknowledgments

In 2008, we sat down with Larry Penny, then the director of the Natural Resources Department for the Town of East Hampton, to explore the idea of writing a "Long Island Fish Book." Larry's duties subsequently prevented him from joining us, but we would like to acknowledge his initiative.

We are particularly indebted to Christopher Paparo, manager of Stony Brook University Marine Science Center and sole proprietor of Fish Guy Photos. Chris is a prominent and productive nature photographer and has contributed all the original high-definition color underwater images and videos.

Special thanks to Rusty Tyler, art director of Cornell Cooperative Extension of Suffolk County, who worked as technical editor on this book. He scanned and enhanced all the historic scientific illustrations and is also responsible for the overall editing details and formatting of the book's interior and cover. Dr. Tobey Curtis, National Marine Fisheries Service (NMFS) fishery management specialist, and Steve Heins, former Mid-Atlantic Fishery Management Council member, provided expert content reviews.

Author HMR owes his interest and career in ichthyology to his university mentors: Tom J. Cade, Syracuse and Cornell Universities; Albert W. Ebeling, University of California (UC) Santa Barbara; and Boyd W. Walker, UC Los Angeles. During the 36 years HMR spent as a member of the Southampton College (Long Island University) faculty, many individuals supported both his teaching and research interests. These individuals include faculty colleagues, Gil Bane, Shirley Baty, Al Berkebile, Christopher Gobler, Thomas Haresign, Maureen Krause, Larry McCormick, Bradley Peterson, William Schutt, Keith Serafy, Sandra Shumway, Alvin Siegel, Joe Warren, and J. Ral Welker; the Marine Station staff, Don Getz, Melanie Meade, and Bruce Ringers; and teaching assistants, Tobey Curtis, Sandra Dumais, Lee Fuiman, and Bruce Mundy. In particular, HMR would like to acknowledge his Tropical Marine Biology teaching partners, Larry Liddle and Stephen Tettelbach.

Further recognition should be given HMR's co-researchers, T. R. Chen, Arthur DeVries, Garth Fletcher, Arthur Goldberg, Rainer Hantsch, Richard Londraville, Franz Luttenberger, Paul March, William Nicol, and David Petzel. Helpful current and former New York State Department of Environmental Conservation fisheries biologists include Chris LaPorta and Byron Young. Tobey Curtis and Greg Metzger served as sources for Long Island coastal shark movements. Aquarists Todd Gardner and Joe Yaiullo served as authorities regarding Long Island subtropical fish visitors. Of special value were the opinions of Todd Gardner and Chris Paparo in reviewing the list of fish species not predicably encountered in New York marine waters.

Michael Maslin generously gave us permission to use his amusing blowfish cartoon from a 2013 issue of *The New Yorker* magazine. Roger Hall allowed us to use the remarkably detailed ink art illustrating the red lionfish.

Special thanks to HMR's wife, Ann Whitman Reisman, for reading manuscript drafts, solving computer problems that confounded her husband, and giving her unfailing support before and during this book project.

Author ECH wants to recognize his wife, Mary Lee, and children, Emerson G. and Emilee, for their support and understanding of his career-long activities and interest in fishes, fishing, fisheries research, and fisheries management as well as his work on this book. ECH credits his interest in fishes to his father and to his undergraduate ichthyology professor, Howard M. Reisman. ECH dedicates this book to his granddaughters Olive and Emerson (Ruby) Hasbrouck, who at a young age already have an appreciation of the importance of the marine environment. Recognition should also be given to ECH's fisheries team coworkers in the Marine Program of Cornell Cooperative Extension (CCE)—John Scotti, Scott Curatolo-Wagemann, Tara McClintock, and Kristin Gerbino, among others. During his nearly 50-year career with NMFS and CCE, ECH has had the opportunity to interact with, work with, and learn from a very diverse group of commercial and recreational fishing industry members, fisheries scientists, and fisheries managers up and down the East Coast. Although way too numerous to list individually, he would like to recognize that all these individuals have increased his knowledge of fisheries and helped to shape his approach to fisheries management. Further, his currently 10 years (and continuing) appointment as the New York Governor's Appointee Commissioner to the Atlantic States Marine Fisheries Commission has given him the ability to develop and implement fisheries management to sustainably manage East Coast species. ECH would also like to acknowledge CCE for providing the opportunity to work on this book as part of his CCE education and outreach initiatives, as well as providing some financial support for the publication of this book.

The authors would also like to thank and acknowledge the two readers who reviewed the manuscript and provided helpful comments and suggested edits. One reviewer is anonymous, the other is Dr. Edward Houde, University of Maryland, Chesapeake Biological Laboratory.

We would also like to acknowledge the helpful advice and editorial services of Susan Specter, Eva Silverfine, and Jacqulyn Teoh of Cornell University Press. We are particularly grateful to Kitty Liu, editorial director of Comstock Publishing, for all her guidance and assistance in shepherding the manuscript through the publishing process.

Abbreviations

ASMFC: Atlantic States Marine Fisheries Commission
CCE: Cornell Cooperative Extension
CGT: Conservation Gear Technology
EEZ: Exclusive economic zone
F: Fishing mortality
FMP: Fishery management plan
HMS: Atlantic Highly Migratory Species Management Division NMFS
ISFMP: Interstate Fisheries Management Program
M: Natural mortality
MAFMC: Mid-Atlantic Fishery Management Council
NEFMC: New England Fishery Management Council
NMFS: National Marine Fisheries Service, also NOAA Fisheries
NOAA: National Oceanic and Atmospheric Administration
NYSDEC: New York State Department of Environmental Conservation
SSB: Spawning stock biomass
Z: Total mortality

A Note on Terminology

Throughout this book we use the term "fisherman" to describe someone engaged in commercial or recreational fishing. The term is meant to be gender neutral as its usage has evolved to be such. Most people engaged in fishing, including women, prefer the use of "fisherman" as opposed to the sterile term "fisher" that has been imposed upon the fishing industry by those not involved in it.

There are various conventions related to the capitalization of the common names of fishes. Throughout this book we use the convention adopted by ASMFC, MAFMC, NEFMC, NMFS, and NYSDEC in all their FMP and stock assessment documents. That convention capitalizes the common name only if it occurs at the beginning of a sentence or if the first name of a two-word common name is a proper noun.

Marine Fishes and Fisheries of New York

Introduction

There are two major goals of this book. The first is to identify New York's marine fishes, with an emphasis on those that are objects of commercial and recreational fisheries or that serve as vital links (keystone species, forage fishes) in the food chain within those waters. Other commonly encountered species with noteworthy biological characteristics are included in order to reveal some of the unique qualities different fishes possess. The second goal is to describe the commercial and recreational fisheries, the methods by which these fisheries are managed, and how sustainable management practices are developed and altered to respond to varying population cycles of fishes, changes in fishing pressures, and changes in environmental conditions. We will also review some of the historical information on catches.

The bulk of the book contains profiles of the fish species to be found in New York's marine waters. Those profiles are organized phylogenetically wherein closely related species are grouped together and families are arranged conventionally from ancestral to more recently derived forms. Each species' profile is preceded by a brief synopsis of the family. Full profiles of major species include information on distribution, size, other common names, anatomical and behavioral distinctions, ecological roles, life history details, and occasional comments regarding food preferences, predators, annual movements, and spawning season. As a result of ongoing ichthyological research, some details regarding number of species in a family, recorded maximum sizes and distribution, and so on, are inevitably being revised. Some species are of significant commercial or recreational value (current or historical). For those, the status of the fishery and its management are discussed. For many other species, only the most basic information and field characteristics are mentioned. The material within the profiles often relies upon primary sources and that literature is cited. There are fewer references for minor species. Some species profiles may be accompanied by an image of a special anatomical feature or a sidebar discussing a topic relevant to that species.

Finally, we have compiled a set of color photos of fish in situ to supplement the historic illustrations; in some cases, we have added a matrix barcode (QR code) that when scanned will direct a reader to an underwater video showing that species in action.

New York has a rich marine fishery history represented by the use of a variety of fishing methods, including those employed by the Native peoples, preindustrial colonials, nineteenth-century fishermen, and modern-day commercial fishermen. Current methods are meant to promote sustainable management of our fisheries. Those practices include the reduction of bycatch and discard mortality, the establishment of artificial reefs, the protection of anadromous and forage fishes by restoring habitats, and the collection of quality data for stock assessments and management. Effective fisheries management is increasingly significant in the face of our present twenty-first-century environmental challenges in the form of climate change, ocean acidification, and energy exploration. Specific regulations are always subject to change, some annually and some more frequently. We have attempted to give the general structure and framework of how individual species are managed without listing the specific fine detail of minimum or maximum sizes, possession or trip limits, open/closed seasons, and other fisheries regulations. These are all subject to frequent change. Specific regulations currently in effect can be found in the links to NYSDEC and NOAA-NMFS websites.

Similarly, as abundance of fish in a population increases or decreases over time for a variety of reasons, the population status may change. Status can refer to abundant, depleted, overfished, not overfished, or whether overfishing is occurring or not. We have provided information on the status of important fish stocks as of the publication of this book. The goal of fisheries management is to try to prevent overfishing and overfished status or to rebuild stocks if those conditions exist.

Any marine fishing activity requires a permit. All marine recreational anglers must enroll in the New York State no-fee recreational marine fishing registry before going fishing in the marine and coastal district waters or when fishing in the Hudson River and its tributaries for "migratory fish of the sea" (e.g., striped bass). In addition to the fishes listed in the New York State registry, most highly migratory species (e.g., tunas, billfishes, and some sharks) also require an additional recreational permit. However, if you are a passenger on a charter or party (head) boat, the boat has a permit to cover you so an individual registration or permit is not needed. To engage in any form of commercial fishing (sale or barter) a state and/or federal permit/license is required. The availability of some of these permits/licenses is restricted, but some are open to all.

We have limited our species coverage to the inland marine waters of New York (harbors, bays, creeks) and those coastal waters within 12 nautical miles of our coastline. This includes Long Island Sound, New York Harbor, and the lower reaches of the Hudson River. Fisheries activity within 3 nautical miles is regulated by the state of New York, whereas federal regulations apply from 3 mi. out to 200 mi. (referred to as the EEZ, the Exclusive Economic Zone). Although some of the species we include may be more common outside of 12 mi., they also occur within this book's coverage area or are important to New York's fisheries. Other than a general summary paragraph on New York's shellfisheries, shellfishes are not the focus of this book.

The Long Island region has the broadest annual water temperature range of any area along the Atlantic coast of North America. Because of that, the diversity of fish species seen within any given year is remarkable. The increase in southern fishes that can be predictably found in Long Island's bays may be associated with the warming of marine waters. Indeed, water temperature in general influences the large-scale distribution pattern of fishes. The recent effects of climate change on the resources and habitat conditions fishes rely upon have caused significant and measurable alterations in their distribution. For example, black sea bass, scup, and summer flounder have shown northern shifts in their distribution, and the shifts of at least the first two were related to increased ocean temperature (Bell et al. 2015).

Briggs and Waldman (2002) presented a comprehensive listing of the fishes reported from the marine waters of New York. The list consisted of 338 species from 114 families of fishes including some deepwater species as well as 12 species of freshwater strays from 5 families. We include only a few deepwater species but do not include any freshwater strays. Further, Briggs and Waldman include more nuanced comments regarding the fishes' relative abundance and documented geographic and seasonal occurrences. In general, we follow the judgments of Briggs and Waldman regarding overall prevalence (abundant to rare) of a species except in those cases in which our more recent personal experiences differ from those authors. Regardless, it is to be expected that over time, most vagile marine organism distributions will be responding to the relatively dramatic changes in ocean water temperatures. We exclude some species that Briggs and Waldman found to be rare with only one local record. Thus, our primary set consists of only 155 species within 76 families. Of these species, 65 are profiled in some detail. The species placed in our secondary list number 89, adding another 25 families and bringing the total number of species we list to 244. Many of these less-common species were noted by Wood et al. (2009) in a study in which 44 warm-water species were collected in Narragansett Bay and Long Island Sound. Those species would have had to pass the south and east ends of Long Island. Further, during a Long Island Sound trawl finfish survey from 1984 to 2008, warm-adapted and subtropical species significantly increased over time (Howell and Auster 2012).

Bean (1901) cataloged the fishes of Long Island and named 241 species (9 rare or of doubtful occurrence). His total included 27 native or introduced freshwater fishes. Therefore, Bean's list contained 214 marine species. This falls short of the 338 species listed by Briggs and Waldman. Even though many common and scientific names have been changed since 1901, Bean's study is remarkable for its time. Also, as a possible measure of changes in the fish fauna, 47 of the species mentioned in this volume are not listed by Bean.

In the early catalog, Bean (1901) listed 13 resident marine species. In most instances, these are fishes that complete all aspects of their natural history in New York waters. This list included the mummichog, striped killifish, rainwater killifish, sheepshead minnow, white perch, black sea bass, cunner, tautog, oyster toadfish, Atlantic tomcod, winter flounder, American eel, and striped bass. In the latter two cases, representatives of these species can be found year-round but some members participate in offshore migrations.

In our current checklist of New York marine fishes, 30 species have been identified as year-round residents. Our additions to Bean's list include the inland silverside, spotfin killifish, fourspine stickleback, threespine stickleback, blackspotted stickleback, ninespine stickleback, northern pipefish, lined seahorse, grubby, silver perch, rock gunnel, naked goby, seaboard goby, smallmouth flounder, windowpane, and hogchoker.

There are several additional studies of the marine ichthyofauna of New York, but these are limited to specific Long Island locations (Orient Point, Peconic/ Gardiners Bay, Fort Pond Bay, Shoreham, Great South Bay, Fire Island Inlet, New York City) or sets of fish-related topics (eggs, young fishes, rare fishes). These contributions are identified in the References subsection Works about Resident and Seasonal Species Found in New York Marine Waters.

Species Accounts

The system of binomial nomenclature, that is, genus and species, was primarily designed by Carolus Linnaeus to assist eighteenth-century scientists to systematically organize living organisms and indicate relationships between them. In the century before, a plant or an animal was named using a polynomial term consisting of multiple words whose origins were Greek or Latin, or both, words and roots. Greek and Latin were the common languages of science at that time, so it was natural for Linnaeus and others to employ those languages when naming the genus and species of an organism. Present-day biologists generally do the same. In many of the species' profiles, where it might be noteworthy, the Greek or Latin origin of the scientific name is included. Scientific names are often descriptive indicating size, form, color, habits, or habitats. Some examples are terms from Latin, such as *borealis* (northern), *brevis* (short), *erectus* (upright), *fuscus* (dark brown), *lateralis* (side), *lineatus* (lined), *maculatus* (spotted), *parvus* (small), *striatus* (striped), and *tenuis* (thin), and from the Greek there is *gaster* (belly), *hippo* (horse), and *rhynchos* (beak).

Common names of fishes might be useful and are more memorable but can be imprecise. First of all, for a species that has an international distribution, the common name would be in different languages. Even when a fish occurs in different parts of the same country, a local common name might not be used elsewhere. For example, two popular sports fishes of New York are fluke and blackfish; however, their recognized common names are summer flounder and tautog, respectively. A consistent common name would reduce confusion but in the words of Shakespeare's Juliet Capulet, "That which we call a rose by another name would smell as sweet." The American Fisheries Society has standardized the common names of North American fishes such that there is a single accept-able common name for each species. Even with this listing, common names can be misleading. The yellow perch (*Perca flavescens*) is not related to the white perch (*Morone americana*). In fact, the white perch is more related to the striped bass (*Morone saxatilis*), which can easily be seen because they are in the same genus. The largemouth bass (*Micropterus salmoides*) is not related to the striped bass at all but is in fact a member of the sunfish family (Centrarchidae), and the sheepshead minnow (*Cyprinodon variegatus*) is unrelated to the freshwater minnow family (Cyprinidae). Winter flounder and summer flounder might be expected to be closely related, but in fact these two flatfishes belong to separate families. So, common names are useful but, as we have seen, not always infor-mative. On the other hand, scientific names reveal close relationships as in the case of the two salt marsh fishes, the mummichog (*Fundulus heteroclitus*) and the striped killifish (*Fundulus majalis*).

ABOUT THE ILLUSTRATIONS

In our species profiles, we elected to use classic scientific illustrations instead of photographs or colored portraits of the fishes. Those illustrations were often created by employees of museums during the late nineteenth century. The most frequently used artist was H. L. Todd, who worked at the U.S. National Mu-seum from approximately 1884 to the early part of the twentieth century. Other illustrators, associated with that and other institutions, included A. H. Bald-win, M. H. Carrington, E. N. Fischer, W. L Hains, C. H. Hudson, M. M. Smith, and M. H. Wagner, among others.

These classic illustrations are remarkable in that they accurately depict the fish's overall shape, scale pattern, fin number and position, fin ray count, tail shape, mouth gape and its relationship to the eyes, and pattern of any body pig-mentation. Those details permitted collectors and scientists to identify the spe-cies. Many of these illustrations were included within part four of Jordan and Evermann's (1886–1900) *The Fishes of North and Middle America: A Descriptive Catalogue of the Species of Fish-like Vertebrates*. The quality and relevance of those images was reaffirmed by their inclusion within the recent editions of the *Fishes of Chesapeake Bay* (Murdy et al. 1997) and the *Fishes of the Gulf of Maine* (Collette and Klein-MacPhee 2002). Bruce Collette, senior systematic zoolo-gist, NMFS, and co-editor of the latter volume, approved of our use of many of the historic, publicly available, illustrations in the 1953 and 2002 editions of *Fishes of the Gulf of Maine*.

The following illustrations are courtesy of the National Museum of Natural History Division of Fishes:

- spotfin killifish, *Fundulus luciae*: catalog # FIN01095; artist Albertus H. Baldwin
- rainwater killifish, *Lucania parva*: catalog # FIN01105; artist W. S. Haines
- Florida pompano, *Trachinotus carolinus*: catalog # FIN03397; the National Museum of Natural History was unable to determine copyright; likely artist H. W. Todd
- pinfish, *Lagodon rhomboides*: catalog # FIN01120, artist unknown
- Atlantic croaker, *Micropogonias undulatus*: catalog # FIN00957; artist unknown

The Academy of Natural Sciences of Drexel University gave permission to use six illustrations of butterflyfishes and surgeonfishes from *Fishes of Bahamas and Adjacent Waters* (Bohlke and Chaplin 1968).

The Royal Ontario Museum allowed us to use the Anker Odum illustration of the blackspotted stickleback.

The illustration of the dentition and pharyngeal plates of the tautog is a product of Shirley Baty's scientific illustration class at Southampton College of Long Island University.

Unless otherwise indicated all illustrations are from sources in the public domain: Hildebrand and Schroeder (1928); Bigelow and Schroeder (1953); Miyake and Hayasi (1972); Colette and Klein-MacPhee (2002); NOAA Photo Library; and the National Museum of Natural History.

INFORMATION SOURCES

All recreational landings data is personal communication from the National Marine Fisheries Service, Fisheries Statistics Division (https://www.fisheries .noaa.gov/data-tools/recreational-fisheries-statistics-queries). The recreational landings represent Type A + B1 (fish that are brought back to the dock that can be identified by trained interviewers + fish that are harvested but identification is by individual anglers). "Personal communication" is how the NMFS wants the website data cited.

Commercial landings from 1950 to present are from NOAA Fisheries Office of Science and Technology, Commercial Landings Query (available at www .fisheries.noaa.gov/foss). Landings prior to 1950 are from McHugh and Williams (1976).

Information for fisheries management sections is from the following:

- ASMFC—management section for each species, http://www.asmfc.org/
- NMFS Atlantic Highly Migratory Species—https://www.fisheries.noaa .gov/topic/atlantic-highly-migratory-species

- NMFS Greater Atlantic Regional Fisheries Office—regulations section for each species, https://www.fisheries.noaa.gov/about/greater-atlantic-regional-fisheries-office
- NYSDEC—recreational regulations and commercial regulations, https://www.dec.ny.gov/things-to-do/saltwater-fishing
- MAFMC—fishery management plans for each species, https://www.mafmc.org/
- NEFMC—management plans for each species, https://www.nefmc.org/

Recreational size records are from NYSDEC, https://www.dec.ny.gov/outdoor/7906.html, and from the International Game Fish Association, https://igfa.org/igfa-world-records-search/.

Literature for biological information is listed with each species.

LAMPREY
Family Petromyzontidae (*lampreas*)

Unlike other fishes living in the waters of New York, the sea lamprey has a mouth but is jawless. Other features including gills arranged in seven pouches on each side, no true bone, and only two semicircular canals rather than three are evidence that this family is a primitive one and is distinctly different from the other major groups of fishes (shark-like fishes and ray-finned fishes). There are eight genera and 34 lamprey species distributed worldwide. In North America there are 20 species of lampreys. With the exception of the sea lamprey and some Pacific lampreys, all of those lampreys occur exclusively in freshwater.

Sea lamprey

Petromyzon marinus (lampetra marina)

East coast of North America, Greenland to northern Gulf of Mexico, west coast of
    Europe to the Mediterranean
Av. adult length = 28 in.

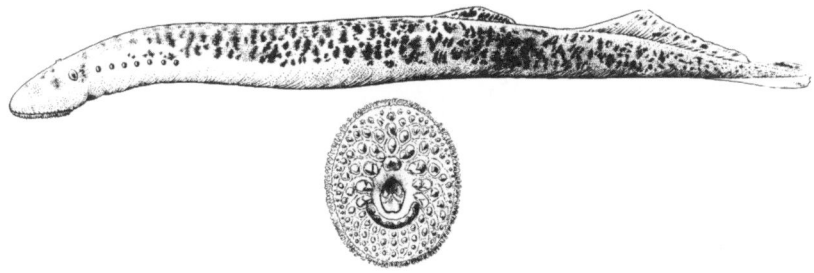

FIELD CHARACTERISTICS. Although the sea lamprey swims in an eel-like fashion, it is distinctly different from the common American eel. They are both elongate, but the eel has one long dorsal fin, as well as a pair of pectoral fins and a conspicuous jaw, whereas the lamprey has two dorsal fins, has no pectoral fin, and is jawless.

ECOLOGY/LIFE HISTORY. The sea lamprey is anadromous and migrates from the sea and spawns in freshwater streams. While moving upstream, a lamprey often uses its suctorial mouth to hold on to rocks or other objects to maintain its position. The name Petromyzon originates from the Greek for "rock sucker." Males generally arrive at the spawning site first, construct a crude nest, and release pheromones (chemical signals) that help attract females. Further, adult lampreys are thought to be attracted to spawning streams by chemicals released by lamprey larvae.

After spawning, adult lampreys die and the larval lampreys (ammocoete larvae) live in the muddy stream banks for about 4–5 years during which they feed by filtering out suspended plankton and detritus. The name "ammocoete" means sand bed. Before the larvae leave fresh water, they undergo a metamorphosis that causes marked changes in their anatomy and physiology. They then migrate back to the sea and assume a parasitic existence until they return to spawn 2–4 years later.

Not all members of the lamprey family are parasitic but the sea lamprey is. As a parasite, *Petromyzon* is very opportunistic and temporarily attaches to the bodies of a wide range of large prey, including whales, sharks, cod, and menhaden. Using their strong suction-like mouths and rasping tongue, parasitic lampreys feed on the host's blood. Most prey survives the attack unless it is small and the attack is prolonged.

The sea lamprey can also be found landlocked in the Great Lakes. The most significant access route was via the man-made canal system built in the nineteenth and twentieth centuries connecting the St. Lawrence River and the Great Lakes. The lake trout and lake whitefish, among other fish populations of the Great Lakes, were threatened by this invasion. For more than 60 years, a lampricide has been used to suppress sea lamprey populations in the Great Lakes. By using lamprey larvicides, the lampreys have been controlled and the threatened fish populations recovered.

LITERATURE. Li et al. 2002

CARTILAGINOUS FISHES

Skates, rays, and sharks are all Chondrichthyes, a class of vertebrates that as the name signifies are "cartilaginous fishes" as opposed to "bony fishes." Cartilage is more flexible and lighter than bone, so some tissues in these cartilaginous fishes that require strength (e.g., teeth and vertebrae) are calcified. Further, such fishes have many additional similarities that distinguish them from bony fishes. In skates, rays, and most sharks, there are five gills on each side rather than four. Placoid scales (aka dermal denticles) differ in composition and form from those scales found in a striped bass, for instance. The term "denticle" accurately suggests that the scales in skates are tooth-like structures with bony material embedded in the skin and covered by enamel. The denticle is sharply pointed and projects backward, and thus skate and shark skins are generally rough. There are no swimbladders in this class of fishes. Swimbladders in bony fishes are used, among other reasons, to achieve neutral buoyancy. In cartilaginous fishes, lift is aided by large fat-filled livers, and sinking is countered by not having a relatively dense bony skeleton.

SAND TIGERS
Family Odontaspididae (*tiburones toro*)

There are only three North American species in this family, the sand tiger, ragged-tooth shark, and the bigeye sand tiger. The gill openings are all in front of the pectoral fin. The dorsal, anal, and pelvic fins are equal in size. The teeth are large, protruding and frequently described as awl-shaped, slender, and sharply pointed. The caudal fin is very asymmetrical, with a long upper lobe and a very short ventral lobe. The oldest sand tiger embryos are oviphagous; that is, while in the uterus the oldest feeds on lesser developed embryos and any eggs.

Sand tiger

Carcharias taurus (tiburon arenero tigre)

Worldwide temperate and tropical, Maine to Texas, absent along the U.S. and
Canadian Pacific coasts

Max. length = 10.3 ft., common = 4.8 ft.

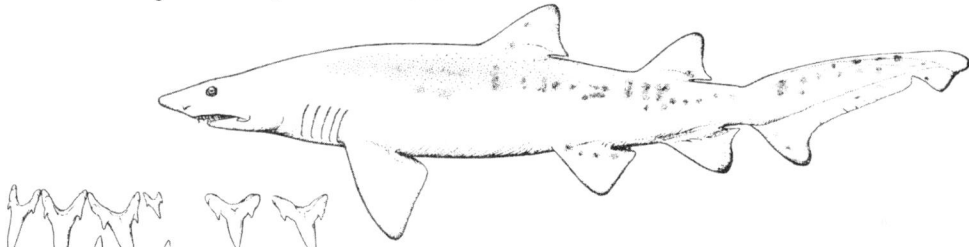

SHARK TEETH

Shark teeth are embedded in connective tissue, unlike teeth in bony fishes that are directly attached to the jaw bone. Shark scales are also only buried in the skin. Further, shark teeth and shark scales (placoid scales) both have a pulp cavity and are covered by an enamel-like substance (vitrodentine). Because of these similarities, it is suggested that shark teeth originate from these scales. Shark scales do not become larger but are added as the shark grows. In bony fishes, scales do grow larger over time so, unlike placoid scales, bony fish scales with annular rings can be used to age the fish.

Shark teeth are constantly being replaced by rows of nonfunctional teeth that grow and move forward to replace those that are lost. The rate at which shark teeth are lost and replaced varies with species, age, and diet, but it is not uncommon for a tooth to be replaced three times a year resulting in thirty to forty thousand teeth lost over a lifetime.

FIELD CHARACTERISTICS. The mouth is large and the teeth are strikingly "fearsome" looking, although this species is an unlikely threat to humans. No lateral keels on the caudal peduncle (the narrow part of the body preceding the caudal fin).

ECOLOGY/LIFE HISTORY. The sand tiger shark, like most sharks, when in motion irrigates its gills by the water that flows through it open mouth, across the gills, and out the gill slits. This form of respiration is called ram ventilation and alternates with active buccal pumping of water from the oral cavity over the gills whenever the shark is not active. Some sharks, for example the sedentary nurse sharks, use only buccal pumping and may also have spiracles behind their eyes that serve as a water intake. Very active sharks that don't stop moving, for example, white and mako sharks, only ram ventilate and do not have spiracles.

LITERATURE. Gilmore et al. 1983

THRESHER SHARKS
Family Alopiidae (*tiburones zorro*)

There are only a few sharks whose profile is easily recognizable, but a thresher shark would be an example. In thresher sharks the curved upper lobe of the caudal fin is as long as the entire body, weighing nearly one-third of the total body weight. The family includes only one genus and three species. Only two species are found in the Atlantic Ocean, the common thresher and the bigeye thresher. The latter does not commonly occur in New York waters.

Common thresher shark

Alopias vulpinus (tiburon zorro)

Other common names: Fox shark
Worldwide warm and temperate oceans, Newfoundland to Florida and South
    America
Max. length = 20 ft.

FIELD CHARACTERISTICS. The genus and species names are Greek and Latin, respectively, for "fox." The long upper lobe of the caudal fin appears to have suggested the long tail in foxes. The common thresher can be distinguished from the bigeye thresher by the two white patches on the ventral surface of the latter and the absence of conspicuously large eyes.

ECOLOGY/LIFE HISTORY. This is a pelagic shark that inhabits the open ocean and can be found from the surface to significant depths although it does come closer to shore when feeding on schools of small fish.

The upper lobe of caudal fin is used to condense and stun schooling prey such as squid and a variety of fishes, for example, menhaden, herring, mackerel, butterfish, and bluefish. Thresher sharks caught on longline hooks as well as on rod and reel are often caught by their tail as they attempted to slap the bait with their tail.

Litters are relatively small, ranging from two to four young. This shark is ovoviviparous; that is, the developing embryos feed on their yolk supply while being carried in the female uterus. After the first one or two embryos develop, the mother continues to produce unfertilized eggs that serve as food for the growing fetuses. Oophagy (eating eggs) by the young also occurs in the lamnid family, that is, white shark, porbeagle, and mako. Upon birth, the young are large, are well nourished, have a good set of teeth, and are experienced cannibals and predators.

LITERATURE. Aalbers et al. 2010; Kim et al. 2013

BASKING SHARKS
Family Cetorhinidae (*tiburones peregrine*)

Only one species is in the family. The gill openings are very large, extending almost to the top of the head. The mouth is large but the teeth are small, although numerous. The most distinctive features are the elongate and hairy gill rakers employed to filter zooplankton that are subsequently ingested. Those gill rakers are actually modified dermal denticles that are routinely shed in the winter.

Basking shark

Cetorhinus maximus (tiburon peregrino)

Worldwide, cool to temperate seas, Newfoundland to Florida
Max. length = 31.8 ft. although larger sizes have been suggested; common length = 25 ft.

FIELD CHARACTERISTICS. This shark will often be seen on the surface, with its snout out of the water, swimming slowly with an open mouth as it follows concentrations of zooplankton, small invertebrates, and fishes.

The basking shark is the second-largest fish. The whale shark at over a maximum of 45 ft. length is the largest fish in the world, but other than a very rare 1935 record, that shark is not a New York area fish although it is found in warmer waters along the North American coast. The whale shark and the megamouth shark complete the short list of the three large filter-feeding sharks.

LITERATURE. Sims and Quayle 1998

MACKEREL SHARKS
Family Lamnidae (*jaquetones*)

This is a prominent shark family but contains only five species. These are found essentially worldwide in tropical to cool temperate waters. Four species occur along the North Atlantic coastline. These include the white shark, two species of makos, and the porbeagle. These species are all good examples of apex predators that as adults have few natural predators and have a significant effect on their ecosystem. The value of an apex predator can be seen when a depletion of such predators results in a disruption of lower food chain relationships. The ability of these and other sharks to bite off chunks of large prey means that, unlike most bony fishes who tend to swallow prey whole, they are not limited by the size of their gape. As a group, these sharks have large triangular teeth, an almost symmetrical, lunate, caudal fin to provide powerful thrust, and lateral keels on their caudal peduncle that serve stability. Further, this family shares with tuna the remarkable ability, for fishes, to maintain a warm body independent of ambient water temperature. This ability has significant consequences in promoting stamina and speed and increasing the rates of neural activity and digestion.

White shark

Carcharodon carcharias (tiburon blanco)

Newfoundland to the Gulf of Mexico in addition to occurring in suitable waters worldwide.
Max. length = 17 ft. male, 19.5 ft. female (~5000 lb.)

FIELD CHARACTERISTICS. The presence of a caudal peduncle with lateral keels is particularly useful for identifying members of the family. However, to distinguish different species close examination of the upper teeth and the relationship between the anal fin to the second dorsal fin is required. In the white shark, the upper teeth are serrated and broad and the anal fin is entirely behind the second dorsal fin. A further distinction is the coloration, that is, the stark difference between the dark back and white belly. This species is related to the largest predatory shark ever known: the extinct 60 ft. *C. megalodon* that lived millions of years ago.

SHARK POPULATION SIZE

In general, marine fishes occur in the millions in our waters. For example, in 2006, 3.8 million striped bass were caught by recreational and commercial fishermen along the North American Atlantic coast. In contrast, a 2011 survey of the number of white sharks off central California estimated that there were, in total, only 219 mature adults (about 13–15 ft. in length) and subadults (about 8–9 ft.) in that region. That number would be less than the population sizes of some other apex predators, for example, killer whales or polar bears. A 2014 re-evaluation of the size of the same white shark population suggested a minimum all-stages population size of at least 2400 individuals in the central California portion of the eastern North Pacific Ocean. The Atlantic northeast and northern California have the highest densities in the United States. Comparable white shark population centers occur off the coasts of South Africa and Australia/New Zealand and to lesser degrees in the northwest Pacific Ocean, Mediterranean, and Central and South America. There is no practical way to actually count the global population of white sharks or any other marine fish species although some tagging studies allow estimations of abundance of many marine fishes. Estimates of the world's white shark population have ranged from 3500 individuals to several times that number, but it is unlikely to be much greater. Among the reasons for this relatively low number, when compared with other fishes, is that a female white shark takes over 30 years to mature, the litter size can be small (2–10), the gestation period is at least 11 months and could be as high as 18 months, and a female spawns only every other year. A 2014 study of white sharks in the northwest Atlantic Ocean suggests that the population is recovering from a decline in the 1970s and 1980s, although these sharks still are sparsely distributed. That study also concluded that during the summer white sharks occur primarily between Massachusetts and New Jersey, and in the winter they shift to Florida and the Gulf of Mexico. Newborns of about 4 ft. in length regularly occur off Long Island, suggesting that those waters serve as a white shark nursery.

LITERATURE. Chapple et al. 2011; Burgess et al. 2014; Curtis et al. 2014; Natanson and Skomal 2015

ECOLOGY/LIFE HISTORY. The vascular system of white sharks has some features that are similar to those found in warm-blooded animals (birds and mammals). These sharks have high hemoglobin and hematocrit levels and a thick muscular ventricle. These properties are used to circulate warm, oxygen-rich blood to the shark's body muscles, stomach, and head. One consequence is the ability to digest seal and other pinniped blubber efficiently. It is suggested by some that white sharks have acquired this physiological adaptation, in part, to compete with killer whales who also favor seals as prey. In addition to preying on live pinnipeds and large fishes, white sharks will scavenge floating dead whales and may gain enough energy from the latter to sustain a shark for several weeks.

Some white sharks have been observed to bite and then release elephant seals, a large marine mammal. Massive blood loss in the prey renders it easier to capture.

The white shark, along with the mako, is among the small number of shark species that have ever been implicated in shark attacks upon humans. Even then, in 2012, a typical year, only 1 of the 47 shark attacks reported in the United States was fatal. It is about a thousand times more likely to be killed by bee stings than by a shark attack.

LITERATURE. Goldman 1997

Shortfin mako

Isurus oxyrinchus (mako)

Newfoundland to Florida, the Gulf of Mexico, South America, and worldwide
Max. length = 12 ft.

FIELD CHARACTERISTICS. As distinct from the white shark, the mako has nonserrated upper teeth and an anal fin that is not entirely behind the second dorsal fin.

ECOLOGY/LIFE HISTORY. Lamnid sharks, like the thresher sharks, are ovoviviparous and oviphagous.

These pelagic sharks are strong swimmers and can make long, 2700 mi., seasonal migrations in a single year. As in all elasmobranchs (sharks, skates, and rays), the shape of their placoid scales helps reduce drag along the fish's body.

Makos are very active and powerful, performing 15 ft. leaps out of the water that require them to attain a significant exit speed (22 mph) in order to perform that maneuver. Makos will feed on other sharks and large fishes. Makos are supreme predators. They have the speed to feed on fast-moving fishes (e.g., swordfish) and the ability to digest them in a stomach whose temperature is 43–46 °F (6–8 °C) above ambient. To add to the digestion process, all sharks have a spiral valve, an internal modification of the lower portion of a shark's intestinal tract. The coiled surfaces of the valve improve the absorption of digested foods.

LITERATURE. Carey and Teal 1969; Carey et al. 1971; Klimley 1999, 2013

SHARK OSMOREGULATION

Most marine fishes are in water that is saltier than their blood. As a result, water tends to be lost from the body via osmosis, and fishes will then drink seawater to replace the water that is lost. The excess salt is excreted from their kidneys and gills. Marine sharks, skates, and rays (elasmobranchs) do not drink seawater but they absorb some seawater and its salt (NaCl) through their gills. That excess sodium chloride is actively secreted by a specialized rectal gland found only in marine elasmobranchs. Their blood is fortified with urea, among other products, so their water loss and gain problems are minimal. The ability to retain and tolerate moderate levels of urea is unique; in other vertebrates, such concentrations of urea are toxic. A dead shark on a boat deck or dock will eventually have a strong ammonia odor due to the breakdown of that urea.

LITERATURE. Marshall and Groswell 2006

Porbeagle

Lamna nasus

Newfoundland to North Carolina and the Southern Ocean
Max. length = 9.7 ft., common = 4.9 ft.

FIELD CHARACTERISTICS. Although the back is a bluish gray, the lower rear portion of the dorsal fin is white. The body appears stout and rounded, and the dorsal fin sits over or just in front of the inner corner of the pectoral fin. Upper and lower jaw teeth are similarly pointed and smooth edged with small cusplets on either side. As in the case of the other sharks in this family, there is an expanded feature of the caudal peduncle that acts as a lateral keel on each side of the peduncle. In the porbeagle there is a furrow (precaudal pit) at the base of the caudal fin both dorsally and ventrally. It has no known function. A caudal pit occurs in many shark species so it is not a unique identifying feature.

HOUND SHARKS
Family Triakidae (*cazones*)

The family of small- to medium-sized sharks is distributed within coastal temperate and tropical waters throughout the Atlantic and Indo-Pacific region. There are nine genera and 38 species of which only 3 occur in the Atlantic (Florida and Gulf smoothhounds and smooth dogfish).

Smooth dogfish

Mustelus canis (cazon dienton)

Other common names: smooth hound, dusky smooth hound
Maine to the Argentinian coast
Max. length = 5 ft., av. adult = 4 ft.

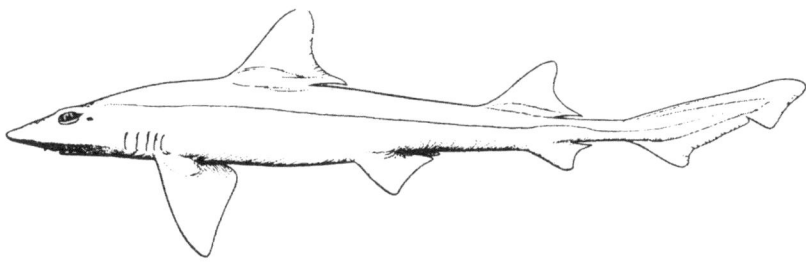

FIELD CHARACTERISTICS. Many sharks may appear similar but the smooth dogfish is most readily distinguished from other sharks because of its modest size, pavement-like teeth, and its nearshore and estuarine occurrence. This bottom-feeding shark uses its flat teeth to crush hard-shelled crustaceans, other invertebrates, and small fishes. Further, this shark often occurs in large groups (packs). Although not unique, this species has a large spiracle behind each eye. A spiracle is a vestigial gill slit and in some benthic or sedentary sharks, and all skates and rays, assists in pumping oxygenated water over the gills.

The smooth dogfish does not really appear like a weasel or a dog but *Mustelus canis* refers to those two animals in Greek. This shark has a uniform grayish upper surface accompanied by a lighter venter. This countershading is a common pattern among fishes. Of particular note is that the smooth dogfish, unlike most other sharks, can temporarily lighten or darken its shade when over white sand or dark sediments.

Another "dogfish," the spiny dogfish, is not in the same family. Although the spiny dogfish is also abundant within New York waters and is modest in size, the differences are distinctive. The most conspicuous differences are the heavy spine in front of each dorsal fin, the absence of an anal fin, and small white spots on the spiny dogfish.

ECOLOGY/LIFE HISTORY. The smooth dogfish is a migratory species and regularly moves from north to south seasonally. New York smooth dogfish mate in our estuaries in the spring and midsummer, then migrate offshore to the south in the fall and winter to regions between Chesapeake Bay and South Carolina, returning to our area in early spring for the birth of the young. Clearly, none of those fish interact with, no less mate with, disparate dogfish populations that extend to Central and South America.

As in all sharks, fertilization is internal. Males use modified pelvic fins called "claspers," which are primarily used to transfer sperm into the female. Mating occurs between May and July. This species has a 10–11-month gestation period. The young develop within the uterus and are nourished via a yolk sac placenta that has folds that are implanted within the uterine wall. This is a clear case of viviparity whereby after full development, independent living young are born. Litter size ranges from 4 to 20 per year per female. In this species, this occurs in late spring within estuaries that serve as nurseries. Smooth dogfish routinely move in packs, and since they often occur in relatively shallow water, they are likely to be the most commonly seen shark within New York waters.

FISHERIES. Smooth dogfish has long been considered a "trash" fish without much commercial demand or recreational interest. In fact, for the most part both commercial and recreational fishermen try to avoid catching smooth dogfish. They are not sporting to catch and often interfere with anglers' efforts to catch more desirable species by quickly taking the bait and reducing the opportunity to catch more desirable fishes. There is no recreational fishery for smooth dogfish.

Except for a small directed fishery and as incidental catch, commercial fishermen avoid smooth dogfish. Dogfish tend to clog fishermen's nets and damage the more desirable catch. The smooth dogfish fishery is largely dependent on the European market where dogfish in various forms have a much higher consumer acceptance than in the United States. Most dogfish landed in the United States are shipped to Europe (and the fins to Asia), so foreign catches, export costs, and currency exchange rates as well as European demand all affect the U.S. fishery. The price paid to U.S. fishermen is low, so they must depend on volume to generate income from the fishery. However, there has been a significant increase in the amount of smooth dogfish landed in New York, and landings have slowly increased over the past 10 years and now average around 250,000 lb./yr. The primary gear used to catch smooth dogfish are trawls and gillnets.

FISHERIES MANAGEMENT. See the general discussion on shark management below.

SHARK SENSORY SYSTEMS

Sharks, in general, are active hunters and as such are always receptive to clues that might lead them to prey. They use all the senses that we would use in addition to some that sense distant vibrations and electric fields. Shark sensory ability varies with the species and environment but, as a rule, shark hearing ability is very acute and has been shown to be able to detect low frequency sound from a great distance, about a mile. As the shark approaches closer to a food source, perhaps to about 1000 ft., it is able to smell chemical stimuli associated with the prey. If the prey is moving erratically, the shark's lateral line may be able to detect that object from a few hundred feet away although the lateral line system usually functions best as a short "distant touch" sense. As the shark comes nearer (about 80 ft. depending on the water's clarity) it will be able to see the prey and attack directly. A final, and very short distance (a few feet), electroreceptive sense is due to a vast array of structures called the ampullae of Lorenzini. These ampullae and their pores are embedded in the skin and are located predominantly in the ventral snout region. The interior of the ampullae is lined by sensory cells that detect electric fields. This last sense is more frequently used to detect prey that might be buried underneath the sediment. This sense has also been suggested to aid in detecting magnetic fields and thus aid the shark during navigation.

LITERATURE. Carey and Teal 1969; Carey et al. 1971; Klimley 1999, 2013; Collins and Whitehead 2004

REQUIEM SHARKS
Family Carcharhinidae (*tiburones gambuso*)

The common name of this family might suggest it refers to their lethal aggressiveness. Some of the family members, for example, the bull shark, oceanic whitetip, and tiger shark, can be so described, and all the family members have formidable blade-like teeth and so are effective predators. However, not every carcharhinid is a menace, and more than likely the name "requiem" is simply derived from the French word for sharks. This is the largest family of sharks with 12 genera and over 50 species. Seventeen of those range to varying degrees along the North American Atlantic coast.

Dusky shark

Carcharhinus obscurus (tiburon gambuso)

Worldwide, warm temperate and tropical from Massachusetts to Florida, Gulf of
    Mexico, Central America
Max. length = 11.3 ft.

FIELD CHARACTERISTICS. The term *obscurus* is Latin for "dusky" and alludes to
the dusky appearance of the pectoral and other fins, although this is not a
unique feature of this species.

Sandbar shark

Carcharhinus plumbeus (tiburo aleta de carton)

Other common names: brown shark
Worldwide distribution but along the western Atlantic coast from Massachusetts to
    Florida, the Gulf of Mexico, and Uruguay; more common north of Cape Hatteras
Max. length = 8 ft.

FIELD CHARACTERISTICS. Color is rarely a distinction in shark identification. The sandbar shark is no exception although its species name *plumbeus* means lead colored in Latin. Its slate gray and brownish hues may have suggested the color of lead when this species was named and identified in 1827.

The first dorsal fin is much larger than the second dorsal fin. The height of the first dorsal fin is equal to the distance from the eye to the third gill slit. The origin of the first dorsal fin is over the pectoral fin base. The lower lobe of the caudal fin is less than half the length of the upper lobe.

ECOLOGY/LIFE HISTORY. The sandbar shark is the most abundant member of its genus in New York and can be encountered nearshore, on sandy bottoms, and in bays and harbors where it feeds on winter flounder, weakfish, and menhaden, among others.

Upper jaw teeth are serrated and a lower set are spiked—well adapted to initially grasp and hold prey, for example, a large fish, and then slice it into edible chunks.

The sandbar shark, like most sharks, when in motion irrigates its gills by the water that flows through its open mouth, across the gills, and out the gill slits. This form of respiration is called ram ventilation and alternates with active buccal pumping of water from the oral cavity over the gills whenever the shark is not as active. Some sharks, for example, the sedentary nurse sharks, use only buccal pumping and may also have spiracles behind their eyes that serve as a water intake. Very active sharks that don't stop moving, for example, white and mako sharks, only ram ventilate and do not have spiracles. Although the sandbar shark is not an obligatory ram ventilator, it too has no spiracles.

This shark is viviparous. Young develop in the uterus of the female. Nourishment is attained through a structure that resembles a placenta. In this case, it is a yolk sac placenta in which the vascular embryonic yolk sac wall fuses with the vascular uterine wall of the mother. Young develop within the uterus for about one year and are born fully formed and at 22 in. in length may weigh 2.2 lb.

Tiger shark

Galeocerdo cuvier (tintorera)

Worldwide, warm temperate and tropical from Massachusetts to Florida, Gulf of
   Mexico, and Brazil
Max. length = 17.8 ft., common = 11 ft.

FIELD CHARACTERISTICS. An alternative species name for this shark was *tigrinus*,
which makes sense considering the pattern of bars and spots on the back
and sides of certain age-class specimens. The current species name, *cuvier*,
is in honor of the noted eighteenth–nineteenth-century French comparative
anatomist and paleontologist, Georges Cuvier.

   In addition to the unique body pattern, the following combination of
features characterized this shark: the snout is blunt, the serrated teeth are
similar in both upper and lower jaws, and the caudal peduncle has a modest
lateral keel on each side.

Blue shark

Prionace glauca (tiburon azul)

Newfoundland to South America and worldwide in warm water
Max. length = 12.5 ft.

FIELD CHARACTERISTICS. As the name suggests, this is a distinctive appearing shark. It has a dark to bright blue (*glauca*) color and a sharply pointed snout (*Prionace*). Further, the body is slender with long, curved pectorals, and the first dorsal fin is set far back. The large caudal fin's sweeping movement thrusts the shark forward. The upper lobe of the caudal fin is twice the length of the lower lobe. In this and other sharks, a strongly asymmetric heterocercal caudal fin also provides some lift and, aided by the position of the stiff and large pectoral fins, helps prevent the shark from sinking. The reason why it would sink is that unlike most bony fishes, sharks, skates, and rays do not have a gas-filled swimbladder.

ECOLOGY/LIFE HISTORY. This may be the most plentiful of large pelagic sharks in the Atlantic and is sometimes seen far out at sea although in summer it may come closer to shore. It is very mobile. Blue sharks tagged off Montauk were recovered two years later off South America and Africa; one tagged shark was recaptured 3630 mi. away on the Liberian coast 9 months after it was initially tagged.

Blue sharks have sharp, finely serrated, triangular teeth. Their diets include a wide range of prey including squid, crabs, gastropods, and such fishes as herring and mackerel. These are viviparous sharks and produce litters ranging from 10 to 50, with some reports of a number exceeding 100.

LITERATURE. Bright 2000

Atlantic sharpnose shark

Rhizoprionodon terraenovae (cazon de ley)

Tropical and subtropical Atlantic both eastern and western coasts, New Brunswick to Florida but uncommon in the north, to the Gulf of Mexico and Central America. The species does not occur in Newfoundland, which curiously is what *terraenovae* is thought to refer.
Max. length = 3.5 ft.

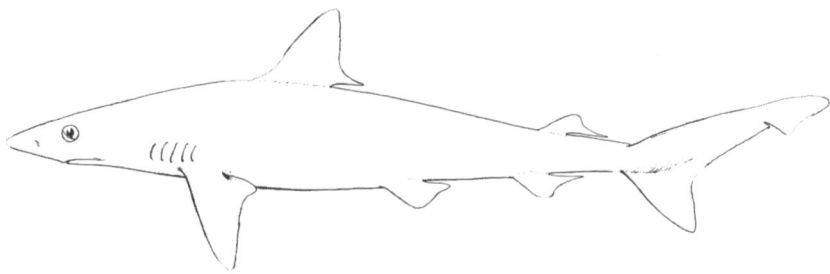

FIELD CHARACTERISTICS. In addition to the long snout that is responsible for the common name for this species, the teeth in both jaws are similarly smooth edged. Further, the second dorsal fin originates posterior to the origin of the anal fin. Adults have scattered white spots on their dorsal surface.

HAMMERHEAD SHARKS
Family Sphyrnidae (*tiburones martillo*)

There are four species of hammerhead sharks that occur along the North American coast but only one predictably occurs in our region. There are eight species distributed worldwide. The species of this family are unmistakable if seen either head on or from above. From either perspective the lateral extensions of the head are apparent with an eye and a nasal opening located far apart on the flattened and expanded head.

Hammerheads are characterized by possessing a distinctive compressed and laterally expanded head (cephalofoil). This structure has several suggested functions including providing greater lift and maneuverability to the shark. Further, the eyes and olfactory organs are located on the extreme outer edge of the head and electroreceptors are broadly distributed along the head's surface, thus enhancing visual, chemical, and electroreception and improving prey acquisition. The New Yorker cartoon suggests that the shark is capable of seeing both eye charts simultaneously, and in fact they can. The scalloped hammerhead shark has a horizontal monocular visual field of 182° on both its left and right side

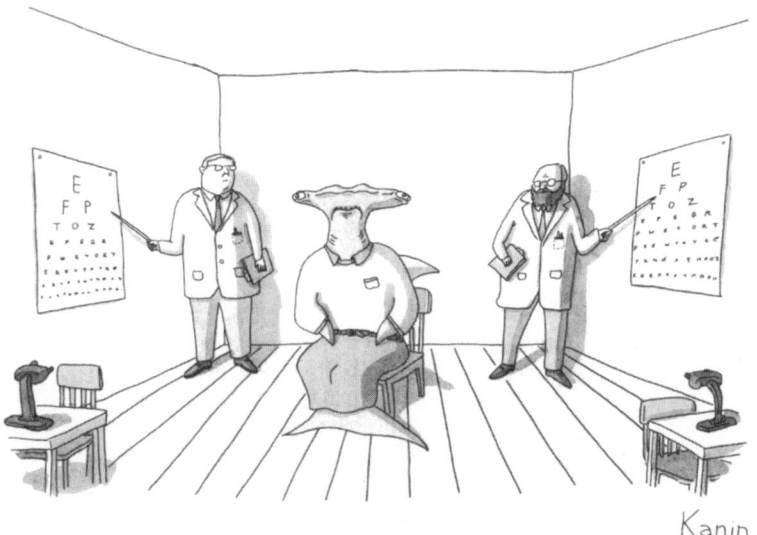

Kanin

and also has 32° of anterior binocular overlap. Binocular vision, when the visual fields of two eyes overlap, allows for improved depth and distance perception and is clearly of importance to visual predators. Binocular overlap increases with the extent of lateral expansion in different hammerhead species. Hammerheads can also move their heads from side to side while swimming and thus can essentially see through 360°. It is suggested that the enhanced vision may have influenced the evolution of hammerhead cephalofoils.

LITERATURE. McComb et al. 2009

Smooth hammerhead

Sphyrna zygaena (cornuda prieta)

Worldwide tropical and temperate seas, Nova Scotia to Florida
Max. length = 11.9 ft.

FIELD CHARACTERISTICS. The head is arched and with a significantly expanded head that is greater than 25% the shark's total length. Teeth in upper and lower jaws are serrated.

LITERATURE. Kaijura et al. 2005

DOGFISH SHARKS
Family Squalidae (*cazones aguijones*)

This family of two genera and over 10 species is characterized by having some of its species with a spine preceding each of the two dorsal fins. Further, all Squalidae lack an anal fin. This family is considered a more recent "derived" shark. Among the approximately 1100 species of cartilaginous fishes (Chondrichthyes)

SHARK CONSERVATION

Three species of hammerhead sharks (scalloped, great, and smooth hammerheads) have been identified by the Convention on International Trade for Endangered Species of Wild Fauna and Flora (CITES) as being threatened, and as such CITES has imposed strict rules regulating their export. Other shark species on the CITES list are either being overfished to unsustainable levels or for other reasons occur in depleted numbers. These species include the white, whale, thresher, and basking sharks, the porbeagle, and the oceanic whitetip as well as shortfin and longfin makos. With the exception of the oceanic whitetip and the great hammerhead, the other shark species occur, although some rarely, within New York offshore waters.

LITERATURE. Shiffman and Hammerschlag 2016

known as sharks (~500), skates, and rays (~600), only the sting rays, with the exception of the spiny dogfish, bear venomous spines, and only the angel sharks, skates, and rays lack an anal fin. In general, this is a family that lives in relatively deep waters.

Spiny dogfish

Squalus acanthias

Other common names: grayfish, spur dog
Greenland to Florida in the western Atlantic, Norway to the Mediterranean in the eastern
    Atlantic; found in the North and South Pacific Ocean but not the Indian Ocean
Av. length= 2–3 ft., max. length = 6.5 ft.

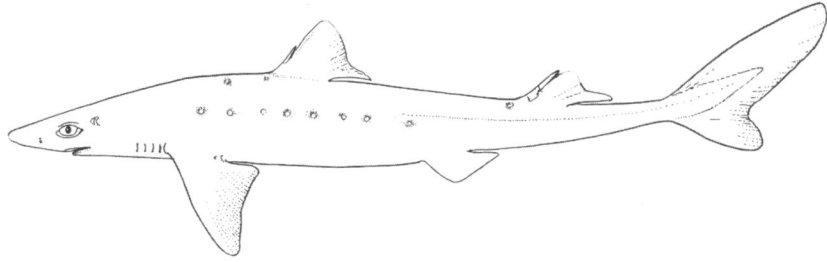

FIELD CHARACTERISTICS. The body tends to have a grayish hue and the sides and back have whitish spots that tend to be reduced in larger specimens. The spines that give this shark its species name (*acanthias*) are venomous and can be painful but are not as potent as stingray spines.

This is a bottom-living species, and as in all such sharks, skates, and rays, the spiracle is relatively large. The spiracle and its pumping mechanism bring in water from above the gill chambers and out the five gill slits as oxygen is being transferred from the water into the shark's blood.

ECOLOGY/LIFE HISTORY. This species is the most abundant shark on the European and North American coasts of the North Atlantic Ocean. Schools can be large with thousands of fish, and those schools can move quickly from one area to another. Spiny dogfish are known to be able to make transatlantic migrations.

Females are ovoviviparous. The young develop internally but gain nutrition from consuming the content of their large yolk sack over a long period of time. Females carry young for between 18 and 22 months, which is among the longest gestation period of any vertebrate, rivaling that of an elephant. Litter size varies but can be from 2 to 11. This species can be found at significant depths, over 2000 ft., but is more coastal than other family members and is even more so when the young are released.

As in many sharks, it has a relatively slow rate of maturation, although this varies. In the northeastern United States the median age is 6.0 years for males and 12.1 for females. Life span also varies geographically. In the Northeast, maximum age is 35–40 years but is longer elsewhere. For the record, some rockfishes, bony fishes of the northeastern Pacific, live for 140 years.

Spiny dogfish appear to be opportunistic feeders with a diet ranging among fin fish, squid, octopus, crustaceans, and soft bodied invertebrates. Because of the abundance of this species and its predation on a wide range of prey, spiny dogfish have a significant role in the ocean's ecology.

In sharks, the liver is enlarged and may weigh from 5% to 25% of the total body weight and occupy a considerable portion of the body cavity. Shark livers are fatty, and the liver oils contribute to the shark's need for buoyancy. Squalene is a major component of the liver oil, but other lipids are used to regulate the buoyancy. Experiments with *S. acanthias* have identified some lesser lipids as being critical in altering the shark's buoyancy. Weights were attached to the dogfish to induce the need to enhance buoyancy. The result was an increase in the amount of low-density diglycerides rather than the denser triglycerides.

LITERATURE. Malins and Barone 1970; Castro 1983; Nammack et al. 1985; Moore et al. 1993; Saunders and McFarlane 1993

FISHERIES. Spiny dogfish was long considered a "trash" fish without much commercial demand or recreational interest. In fact, for the most part both commercial and recreational fishermen try to avoid catching dogfish. They are not sporting to catch and often interfere with anglers' efforts to catch more desirable species by quickly taking the bait and reducing the opportunity to catch more desirable fishes.

Except for a small directed fishery in New York and as incidental catch, New York commercial fishermen avoid dogfish. Dogfish tend to clog fishermen's nets and damage the more desirable catch. However, in some other Mid-Atlantic and New England states there is a large directed fishery for spiny dogfish. The spiny dogfish fishery is largely dependent on the European market, where dogfish in various forms have a much higher consumer acceptance than in the United States. Most dogfish landed in the United States are shipped to Europe (and the fins to Asia), so foreign catches, export costs, and currency exchange rates as well as European demand all affect the U.S. fishery. The price paid to U.S. fishermen is low, so they must depend on volume to generate income from the fishery. In the past, dogfish were also harvested for health-related products derived from their liver. These products include squalamine (an antimicrobial sterol), omega-3 fatty acids, and vitamins A, D, and E. However, since most of these can now be made synthetically, there is little demand for dogfish liver for these compounds.

Spiny dogfish has been a relatively small component of New York's commercial fishery since the 1940s. Peak landings of 1.9 million pounds occurred in New York in 2000. The catch then declined as management measures were put into place. Since 2009 catches in New York have averaged between 200,000 and 400,000 lb., depending on quota and market demand. The primary gear used to catch dogfish are trawls and gillnets.

MANAGEMENT. Spiny dogfish is managed jointly by the MAFMC, NEFMC, NMFS, and ASMFC through the Spiney Dogfish Fishery Management Plan (FMP) (MAFMC 1999; ASMFC 2002). After fishermen were encouraged to target spiny dogfish as an underutilized species, the spiny dogfish resource was declared overfished in 1998, which led to the MAFMC and NEFMC joint FMP implemented in federal waters in 2000 and the ASMFC FMP implemented in 2003 for state waters. Spiny dogfish is sustainably managed, and the resource was declared rebuilt in 2010. Additionally, the resource presently is neither overfished nor is overfishing occurring (MAFMC 2018).

The spiny dogfish commercial fishery has been managed under a quota system since 2004. The management plan subdivides the total quota into a northern region (Maine through Connecticut) share of 58% and a southern region share of 42% with state-by-state subquotas. Of the 42% southern region share, New York receives 2.707%. The NYSDEC then sets trip limits and possible closures for New York fishermen. There is no minimum size, but the market prefers larger fish. There are no size, season, or bag limits for the recreational fishery in New York.

ALL TACKLE WORLD RECORD: 15 lb. 12 oz.

NEW YORK RECORD: There is no New York record for spiny dogfish

SHARK MANAGEMENT

Owing to their vast migration habits along the Atlantic coast and throughout the Atlantic Ocean, all sharks (with the exception of spiny dogfish) are managed jointly by the ASMFC and the NMFS Atlantic Highly Migratory Species Management Division (HMS). The MAFMC and the NEFMC are not involved. The United States has implemented some of the strongest shark management measures worldwide in order to manage shark species sustainably. Increased fishing effort on sharks, both commercial and recreational, led to overfishing on many species of sharks. Many shark stocks in the Atlantic continue to be depleted and are under rebuilding plans.

The HMS began to develop management plans for sharks in the 1990s. All sharks are now federally managed with other highly migratory pelagic species under the Consolidated Atlantic Highly Migratory Species Management Plan and Amendments. The Shark Conservation Act of 2010 also provides additional protection and management measures for sharks, including protecting sharks from illegal, unreported, and unregulated fishing activities. The ASMFC manages 40 different species of coastal sharks (except spiny dogfish) through

CATEGORIES OF SHARKS FOR MANAGEMENT PURPOSES

MANAGEMENT GROUP	SPECIES WITHIN GROUP
Prohibited	sand tiger, bigeye sand tiger, whale, basking, white, dusky, bignose, Galapagos, night, Caribbean reef, narrowtooth, Caribbean sharpnose, smalltail, Atlantic angel, longfin mako, shortfin mako, bigeye thresher, sevengill, sixgill, and bigeye sixgill
Small coastal	Atlantic sharpnose, finetooth, bonnethead, and blacknose
Blacknose	blacknose
Large coastal	silky, tiger, blacktip, spinner, bull, lemon, nurse, sandbar, scalloped hammerhead, great hammerhead, and smooth hammerhead
Pelagic	porbeagle, common thresher, oceanic whitetip, and blue
Smoothhound	Smooth dogfish and Florida smoothhound

Source: NMFS-HMS (https://www.fisheries.noaa.gov/atlantic-highly-migratory-species /atlantic-highly-migratory-species-fishery-compliance-guides).

the Interstate FMP for Atlantic Coastal Sharks. The ASMFC plan follows and complements federal management actions in order to provide for management of sharks in state coastal waters.

The HMS and ASMFC shark management plans combine various species of sharks (except spiney dogfish) into shark management groups to facilitate management measures. These management groups and the species they contain are detailed in the accompanying table. Retention is allowed (within regulations) except for the Prohibited Group. The NYSDEC has implemented commercial and recreational regulations in New York waters consistent with the ASMFC plan.

Commercial fishermen are restricted by quotas, trip limits, and limited access permits. Recreational fishermen are restricted by bag limits and minimum sizes. Throughout the Atlantic, fishermen are prohibited from keeping 19 species of sharks, including those that occur in New York waters (i.e., sand tiger, basking, white, and dusky sharks).

In 2010 the United States enacted a national ban on shark finning that prohibits any person from engaging in shark finning and possessing shark fins without the corresponding carcasses (except for smooth dogfish). Finning is the practice of removing the fins from a shark and discarding the remainder of the shark.

TORPEDO ELECTRIC RAYS
Family Torpedinidae (*torpedoes*)

There are two families of electric rays in the world, the torpedinid electric rays and the narcinid electric rays. Combined there are 59 species, but only 1 species occurs in the area around Long Island. The electric organs are derived from striated branchial muscles in the head region and can generate voltages as high as 200 V depending upon the size of the specimen. This discharge serves to stun prey as well as discourage predators.

Atlantic torpedo

Torpedo nobiliana (torpedo del Atlantico)

Temperate and tropical Atlantic, Nova Scotia to Gulf of Mexico, most between Cape
Cod and North Carolina, eastern Atlantic and Mediterranean
Max. length = 5.8 ft.

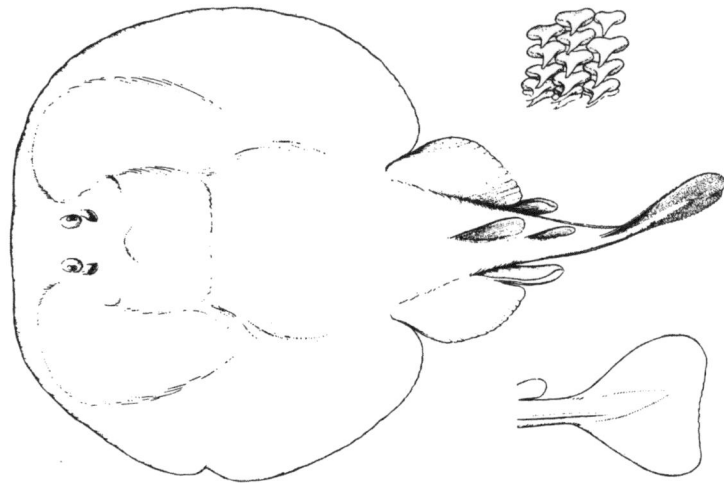

FIELD CHARACTERISTICS. Unlike stingrays and cownose rays, a torpedo has
neither long tail nor venomous spine. The paddle-shaped caudal fin is
distinctive as well as the kidney-shaped patterned surfaces (the electric organs)
occupying considerable areas on both sides of the round pectoral disk.

LITERATURE. Feng 1991

SKATES
Family Rajidae (*rayas*)

In a 2006 enumeration of the world's fishes, the skate family was cited as having 26 genera and about 218 species. As a result of ongoing taxonomic studies
these numbers are likely larger now. Regardless, the number of skate species is
easily more numerous than in any other family of cartilaginous fishes. Skates
are broadly distributed worldwide, ranging from polar regions to the tropics.
As a group they are more common in cooler waters in contrast to rays that are
more numerous in warmer temperate waters. There are 14 skates along the East
Coast of the United States.

Skates and rays are somewhat similar in their overall form, having large spiracles behind the eyes, and assume a general benthic habitat. The absence of a spine on the tail and the presence of prepelvic spurs in the skate aid in distinguishing skates from rays. Prepelvic spurs are the small additional anterior lobes of the pelvic fin. Unlike rays, skates have caudal and dorsal fins.

Skates are benthic fishes and thus find their food on or near the bottom. That food consists of crustaceans, shellfish, worms, other invertebrates, and a variety of small fishes. The skates move along by pectoral fin undulations and gentle pushes as their prepelvic spurs come in contact with the bottom. Predators include sharks, seals, and large fishes such as cod, bluefish, and summer flounder.

The strongly compressed shape, ventral position of the mouth, and bottom-living habit of skates makes the routine pumping of water through their mouth and over the gills difficult. Instead, water is pumped from above through the large spiracles located behind each eye. That oxygen-rich water flow then passes over the gills and out each gill slit.

Barndoor skate

Dipturus laevis

Other common names: sharpnose skate
Newfoundland to North Carolina
Max. length = 5.0 ft.

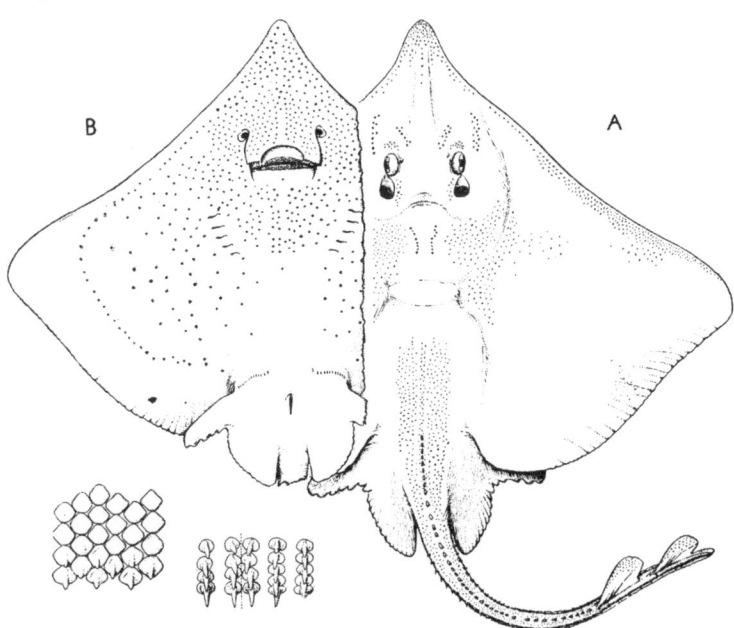

FIELD CHARACTERISTICS. All the commonly observed skates in the region have brown dorsal surfaces with varyingly distinct dark spots or lines. In the barndoor skate, large thorny scales are absent along the middorsal region (A) of the disk. The snout is relatively pointed so that a straight line from the tip of the snout to the tip of the pectoral fin does not cross the disk. Dermal denticles are on the snout and around the eyes. On the ventral surface (B), there are dark dots over the pores of the ampullary and lateral line system. In this and other skates, the shape of the snout, presence of denticles at various locations, and the characteristics of the teeth will be influenced by the size and sex of the individual.

LITERATURE. Gedanke et al. 2005

Little skate

Leucoraja erinacea

Other common names: common skate, summer skate, hedgehog skate
Newfoundland to North Carolina
Max. length = 1.7 ft.

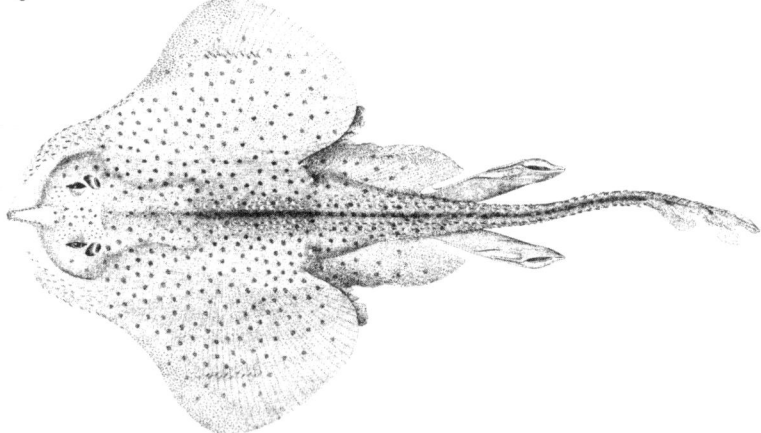

FIELD CHARACTERISTICS. The little skate is appropriately named since it is approximately one-third the size of the equally common winter skate. The species name, *erinacea*, means "like a hedgehog" in Latin. That scientific name was likely inspired by the thorny denticles on the pectoral fins and tail region.

Little skate vary in the distribution of thorns on the surface, size and shape of any dark markings on the dorsal surface, and shape of the disc and snout. The little skate has distinct small round dark spots, an obtuse snout extending beyond the anterior margin of the pectoral fin, less than 21 thorns along the midline of the tail, and a dorsal surface that is relatively free of denticles except for some distinct patches. One of the challenges in identifying some skate species is that some of the denticle patterns change with age and growth.

ECOLOGY/LIFE HISTORY. Reproduction in the little skate is representative of skates in general. Fertilization is internal, employing the specialized pelvic fins of the males to assist in the introduction of sperm into the female oviducts. Once fertilized an egg is encapsulated in a tough rectangular egg case made of collagen and keratin. The eggs are deposited on the sea floor, where the skate develops outside the body of the female, the general mode of reproduction called oviparity. In the little skate, the egg case ("mermaid's purse") has a horny extension and tendrils on each corner. The tendrils, when entangled in bottom vegetation, help to anchor the egg case. The egg case horns are also avenues for water exchange between the environment and the developing embryo. Within 6 to 9 months, the embryo's yolk supply is diminished, and when completely consumed, the fully independent skate hatches.

Skates are among the few fish families that have electric organs. Using their electric organs located in their tail section, the little skate can use weak electric discharges (1.5 V) to communicate during social and reproductive interactions. Unlike the strongly electric ray, *Torpedo*, their electrogenic ability is not used for defense or to capture prey.

LITERATURE. Bratton and Ayers 1987; Summers and Koob 1996

Winter skate

Leucoraja ocellata

Other common names: big skate, eyed skate
Newfoundland to South Carolina
Max. length = 5.0 ft.

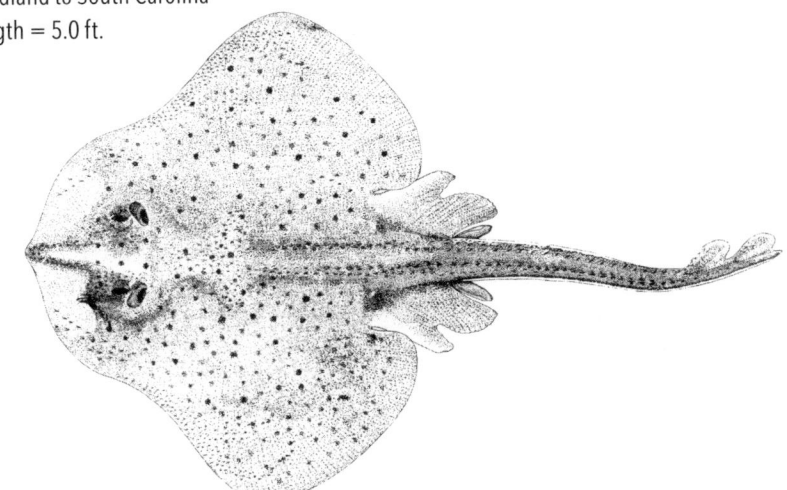

FIELD CHARACTERISTICS. The snout is blunt somewhat as in the little skate. Often, a relatively unique white eye spot occurs on each pectoral fin along with dark round spots. There are three or more rows of thorns along the midline of the tail from the middle of the back to the origin of the first dorsal fin. The latter feature is only true in small (<27 in.) winter skate. The presence, form, and distribution of thorns and dermal denticles vary as the skate approach maturity.

Clearnose skate

Raja eglanteria (raya naricita)

Massachusetts to Florida and Texas
Max. length = 3.0 ft.

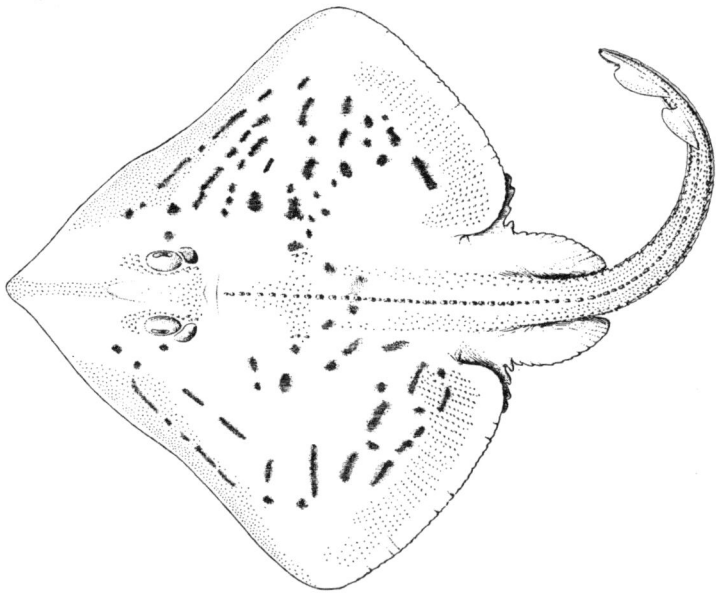

FIELD CHARACTERISTICS. The blotches and markings on the dorsal surface are much more bar-like than in other skates. The snout is slightly acute. The dorsal and lateral surfaces of the tail each have a row of thorns. Although other skates may have rostra that are somewhat translucent, the clearnose skate has large triangular areas on either side of the snout's midline ending just anterior to the pectoral fin rays.

WHIPTAIL STINGRAYS
Family Dasyatidae (*rayas latigo*)

Stingrays are, along with skates and sharks, cartilaginous rather than bony fishes. The whiptail stingray family is large with six genera and 68 species distributed worldwide, mostly in warm temperate and tropical waters. In Southeast Asia, some members of this family occur in brackish and freshwater. Only five species are found within North American Atlantic waters. The tail is very whiplike and is longer than the breadth of the disc. When undulated, the disc, an expansion of the pectoral fins, is the primary mode of locomotion. There is no anal or caudal fin. The most distinctive features of stingrays are the serrated spines. The one or two spines are located relatively near the base of the tail. Glandular tissue near the tip of the spines produces venom. When a stingray is attacked, the tail bends backwards and a spine is simultaneously thrust upward to puncture the body of an attacker. The venomous tissue in the spine is ruptured and elicits a painful response.

Roughtail stingray

Dasyatis centroura

Massachusetts to Florida, Gulf of Mexico, and Mediterranean
Disc width over 6 ft., total length to tail tip is near 10 ft.

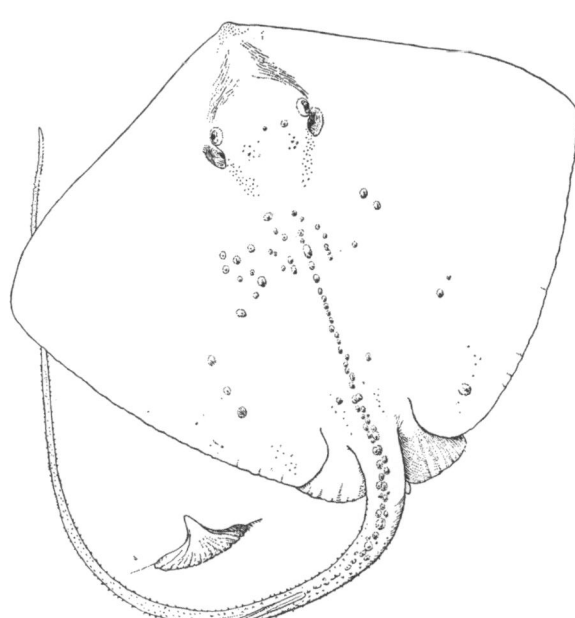

FIELD CHARACTERISTICS. This is the largest stingray in North America. The largest marine stingray in the world is the short-tail stingray of southern Africa to New Zealand (6.9 ft. wide and 14 ft. long); it is also a member of the whip-tail stingray family. Local rays and skates are often confused because both have flattened bodies, expanded pectoral fins, and a large spiracle behind each eye, are bottom living, and have eggs fertilized internally. The latter ability requires males to have modified pelvic fins. Rays, unlike skates, generally have smooth bodies except, as in the roughtail stingray, there are some patches of thorny denticles on the disc and tail. The long, serrated spine in stingrays may have arisen through the enlargement of certain denticles. These spines may be shed annually and replaced.

ECOLOGY/LIFE HISTORY. Despite the superficial similarities, rays and skates are significantly different in how young develop. In rays, the fertilized eggs are retained within the female where "milk-producing" uterine tissue extends into the mouth and gill chambers. Live young are carried internally (viviparity) until birth after a gestation period of from 9 to 11 months.

The mouth in this benthic species is located on the underside of the disc. The jaws have multiple rows of blunt teeth primarily adapted to feed on hard-shelled invertebrates, although small fish are taken as well. In general, sting rays have few predators except for some large sharks and marine mammals. The most effective defense stingrays have against predators is the ray's ability to bury in the soft sediment and thus escape detection.

EAGLE RAYS AND MANTAS
Family Myliobatidae (*mantas y aguilas marinas*)

Myliobatidae occur worldwide along warm temperate and tropical coasts and includes 37 species. The seven species of cownose rays are often placed in their own family of Rhinopteridae. The tails are long and whiplike with a serrated spine at their base. There is no caudal fin. The head is elevated above the disk and the eyes and the spiracles are located on the side of the head rather than on top of the head. The dentition is in the form of a series of plate-like teeth and is used to crush mollusks and crustaceans. The loss of sharks that normally prey on cownose rays was once thought to have the cascading effect of reducing some bay scallop populations along the East Coast. However, a re-examination of data suggests declines in large coastal sharks did not coincide with rapid increases in cownose ray abundance nor did that increase coincide with declines in those scallops.

Cownose ray

Rhinoptera bonasus (gavilan cubanito)

Massachusetts to Florida, Gulf of Mexico to Argentina
Disk width max. = 3 ft.

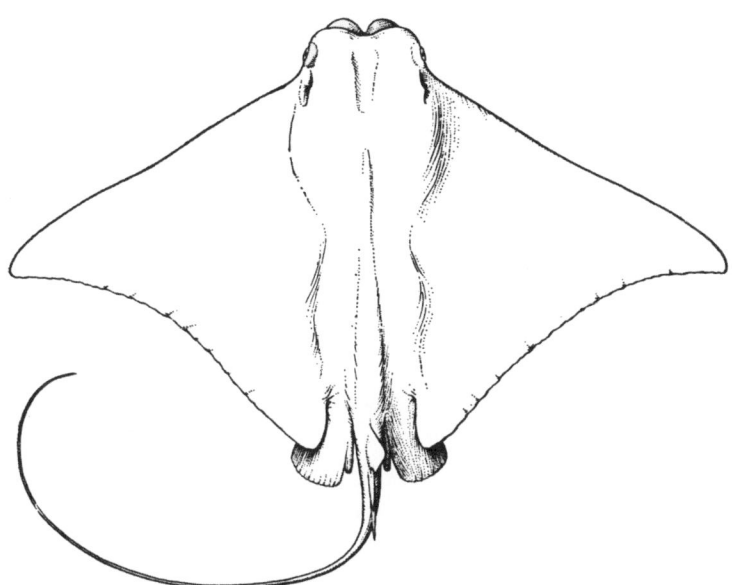

FIELD CHARACTERISTICS. The snout is bilobed and indented. The dorsal fin is located between the pelvic fins rather than posterior to the pelvic fins. The trapezoidal shaped pectoral disk is broader than long.

LITERATURE. Myers et al. 2007; Enzor et al. 2011; Grubbs et al. 2016

RAY-FINNED BONY FISHES

STURGEONS
Family Acipenseridae (*esturiones*)

There are 25 species of sturgeons in the Northern Hemisphere. Eight of those species are in North America. Some are restricted to freshwater, others are anadromous. Two species, the Atlantic and the shortnose sturgeons, are in our region.

This family is considered an ancient family based upon the fossil record and the anatomy of present species. Family members have not changed very much from those extinct species of over 100 million years ago. Common ancestral features include heavy bony scutes instead of conventional scales, a cartilaginous skeleton, and a vertebral column that turns upward (heterocercal tail), often making the caudal fin's dorsal lobe larger. Both sturgeons and sharks have cartilaginous skeletons although they are not closely related. Both of these fish groups had their origins from early bony fish ancestors and have secondarily lost the bony skeleton, replacing it with cartilage that has some ossification. Other features that help characterize the family include a benthic life style, a downward-directed mouth, and a set of four barbels under the snout. The family includes four genera and 25 species. Many attain considerable size, for example, the 18 ft., 4400 lb. beluga from the Caspian, Black, and Adriatic Seas.

Shortnose sturgeon

Acipenser brevirostrum

New Brunswick to Florida
Max. length = 4.6 ft.

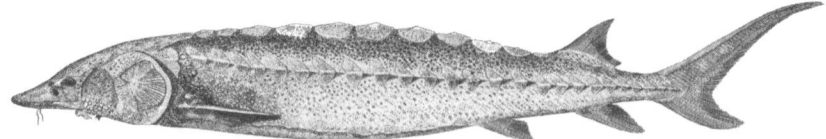

FIELD CHARACTERISTICS. This species is appropriately named *brevirostrum*, Latin for short snout. In addition to the length of the snout and width of the mouth, the shortnose sturgeon differs from the Atlantic sturgeon in the absence of bony plates between the anal fin and lateral scutes in the shortnose sturgeon. Further, if the abdominal cavity is exposed, the intestines and the peritoneum (membrane lining the cavity) are dark and not pale in the shortnose sturgeon.

Atlantic sturgeon

Acipenser oxyrinchus (esturion del Atlantico)

Labrador to Florida and the Mississippi River
Max. length = 16 ft.

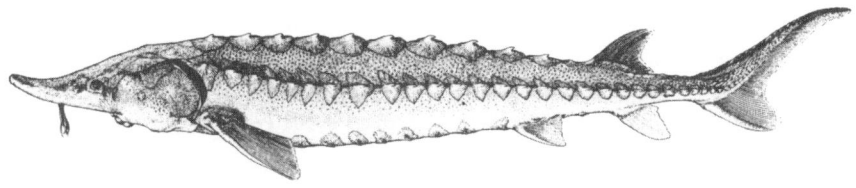

FIELD CHARACTERISTICS. Atlantic and shortnose sturgeons differ in the number of rays in their anal fin, the number of gill rakers, and the number of scutes. Countable anatomical features are called meristics. Measurable features (morphometrics) are also useful and, in fact, the most apparent difference between these two species is the width of their mouths. The Atlantic sturgeon has a mouth width slightly more than one-half the distance between the eyes whereas the shortnose sturgeon has a wider mouth whose width is approximately three-fifths the distance between the eyes.

ECOLOGY/LIFE HISTORY. This is an anadromous fish, and mature adults move toward freshwater in the spring. In large rivers, spawning occurs at the interface of the salt front and the freshwater source. Eggs are broadcast and after fertilization receive no parental care. Females may only spawn every 3–5 years. The spawning period lasts until midsummer. Eventually, the adults move back out to sea. In long-lived fishes, sexual maturity is often delayed for several years until individuals can age. Some sturgeons can be 40 years old before attaining maturity.

Sturgeons are benthic feeders. Their toothless mouths are protrusible and flexible, and their mode of feeding has been described as being like a vacuum cleaner moving slowly along the bottom and sucking up worms, small crustaceans, and small fishes. This process is assisted by the set of chemosensitive barbels in front of the ventrally located mouth. The "inferior" mouth itself is located well below the snout.

LITERATURE. ASMFC 1990; Billard and Lecointre 2000

FISHERIES. Atlantic sturgeon has been harvested in New York since colonial times. As with many other nearshore fishes, Atlantic sturgeon experienced its highest landings in the late 1800s and early 1900s when landings were around 40,000 lb./yr. Peak landings of 428,000 pounds occurred at the end of the nineteenth century. Much of that catch was taken from the Hudson River, a major spawning river for sturgeon along the Atlantic coast. Fish were caught by cotton gillnets and set-lines. Sturgeon have been harvested for their meat as well as their roe, which makes excellent caviar. In the late 1800s much of the Hudson River-caught sturgeon went to Albany and became known as "Albany beef." After that, landings dropped to a few thousand pounds per year. There was a resurgence of sturgeon landings in New York during the late 1980s and early 1990s from the inshore ocean as bycatch in fisheries directed at other species with trawls and modern monofilament gillnets. Landings were 40,000 to 60,000 lb./yr. during this time, with a high of 83,000 lb. landed in 1991. Landings dropped off precipitously after that because of management regulations and a moratorium on landings to protect the species (see below). Landings have been zero since the late 1990s. Even before the moratorium there was not a recreational fishery for sturgeon.

MANAGEMENT. The East Coast Atlantic sturgeon stock comprises five distinct population segments, including one identified as New York Bight. All five population segments are depleted. They are considered "depleted" rather than "overfished" because of the many factors that have led to low abundance, including fishing, habitat loss, and water quality (ASMFC 2017b). Atlantic sturgeon is managed by the ASMFC and the NYSDEC through the Atlantic sturgeon FMP (ASMFC 1990) implemented in 1990 to reduce fishing mortality and help rebuild the stock. Sturgeon is also managed by the NMFS through the Endangered Species Act (ESA). As part of the FMP a coast-wide moratorium was implemented in 1998. There has been no harvest since then, but there continues to be incidental bycatch mortality. In 2012 four of the five population segments (including New York Bight) were determined to be endangered under the ESA, and the fifth was determined to be threatened. In addition to the harvest moratorium, the ESA designation provides for additional protection for the species.

FRESHWATER EEL
Family Anguillidae (*anguilas de rio*)

There is only one genus (*Anguilla*) within this family. This genus has 15 species distributed throughout the world's coastal regions and oceans. Only one species occurs in North America, the American eel, *Anguilla rostrata*. Worldwide, there are many other distinctive eel families containing over 700 species. Anguillidae is the single family whose members spend most of their life history in freshwater or estuaries, only going to the open ocean to spawn at the end of their life cycle.

All the eels, regardless of their family affiliation, share common morphological and developmental features. Eels have no pelvic fins, pelvic girdle, or conspicuous scales.

American eel

Anguilla rostrata (anguilla americana)

Other common names: Atlantic eel, silver eel, freshwater eel, elver (juvenile)
Greenland southward to the coast of Brazil; in North American inland waters (Great Lakes, Mississippi River)
Max. length = adult female (4 ft.), male (3 ft.)

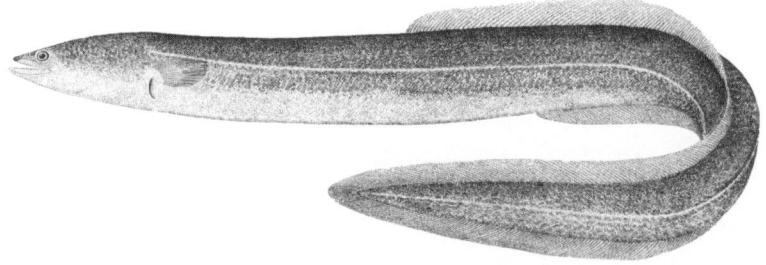

FIELD CHARACTERISTICS. The species name *rostrata* is Latin for "long nosed" and refers to the eel's general shape. Eels have a small gill opening in front of each pectoral fin. The long dorsal fin is continuous with the caudal and anal fin.

ECOLOGY/LIFE HISTORY. As in all fishes, most of the eel's biology is directed toward surviving to reproductive condition and reproducing successfully. In the American eel, these objectives have led to a notable set of morphological, physiological, and behavioral adaptations.

All populations of the American eel are the result of spawning at great depths in the Sargasso Sea, an area east of the Bahamas and south of Bermuda. After several months, the planktonic leptocephalus larvae approach the North American coastline; they are transformed into transparent eel-shaped "glass eels" and then into grayish green pigmented "elvers" as they arrive in the estuaries. Within a few months, the elvers grow and become yellowish, greenish, or olive-brown young eels that are commonly seen in our bays and harbors. While growing to large breeding adults, the eels can be found in fresh, brackish, or saltwater. They feed mostly at night and use their keen sense of smell to locate fishes, crabs, and mollusks. Their food can be either living or dead. (Eel traps commonly use horseshoe crab halves as bait.)

After a varying number of years (5–20), the eels, which now appear silvery, undergo more extensive morphological and physiological changes in preparation for their reproductive migration back to their natal deep-sea spawning area. To be able to swim great distances (1000 mi. from New York) and survive pressures and living conditions found at depths that may exceed 1000 ft., the eels have stored fat, increased the volume of red musculature, fortified the wall of their swimbladder, modified the retinal pigments in their enlarged eyes, and changed their method of osmoregulation. These extreme modifications are completed during the months preceding their seaward migration. They do not eat during the migration, and their digestive tract and olfactory system degenerate. Migrating eels may use a geomagnetic sense to direct their movement and stop when they come upon a thermal front marking the general spawning location. Maturing Nova Scotian eels were equipped with pop-up satellite archival tags. One eel, after swimming eastward across the continental shelf, swam straight south. In total, the tag recorded a migration of 1440 mi. to the northern limit of the Sargasso Sea spawning site. As of this writing, no adult eel has ever been collected from the Sargasso Sea, and the adults die after spawning; however, the millions of young eels that arrive at our coastline attest to the success of the spawners.

Considering the wide range of geographic locations from which the parents originated along the coast of North America, it is unlikely that a young eel will return to the same location from which the parents came. Fishes, like anguillids, who live most of their lives in freshwater and migrate down to the sea to reproduce are called catadromous. There are no other local catadromous fishes. The reverse migratory pattern is known as anadromy. The best-known examples of anadromous fishes are various salmon species, and in New York anadromy can be most commonly observed during the spring spawning migration of alewives.

LITERATURE. Kleckner 1980; Beguer-Pon et al. 2015

FISHERIES. The American eel has been harvested in New York since colonial times. The New York commercial fishery was at its heights during the 1930s through 1950s. Peak landings of 375,000 lb. occurred in 1951. New York landings fell to near zero levels in the late 1990s and have averaged around 30,000 lb. since 2011. Eels are harvested for both food and bait, but most of the demand is as bait (primarily for striped bass). There is also a very high export demand for the small elver stage of eels ("glass eel") to be used in aquaculture operations, but it is not legal to harvest glass eels in many states, including New York. Eels are harvested primarily with specially designed eel pots. Some are also harvested with pound nets. They have also historically been harvested with specially designed multiprong eel spears fished through the ice in winter. Yellow eel (juvenile) is the primary life stage harvested by fisheries. There is a very limited harvest of the silver eel (adult) life stage.

Although not a very popular recreationally caught species, there is a recreational fishery for eels in New York. Recreational catches have varied widely and have ranged from 0 lb. in some years to over 200,000 lb. in 1983. Since 2000 recreational landings have ranged from 0 to 72,000 lb. However, recreational catch estimates of eels are very imprecise.

ALL TACKLE WORD RECORD: 9 lb. 4 oz., New Jersey

NEW YORK RECORD: None

MANAGEMENT. The American eel is managed within 3 mi. of the coast and in inland waters by the ASMFC and the NYSDEC through the American Eel FMP implemented in 1999 to reduce fishing mortality and help rebuild the stock (ASMFC 1999). The American eel population in U.S. waters is depleted. The abundance of eels in New York and along the East Coast has been significantly reduced since the late 1980s. This reduction has been caused by fishing as well as a range of environmental impacts, including dams that prevent and/or restrict elvers and juveniles to ascend into freshwater for that portion of their life cycle. Impingement in the cooling water systems of electric generating facilities has also affected abundance.

American eel is a difficult species to manage due to a number of challenges including (a) it has a variety of life stages with a wide range of habitat needs; (b) the various life stages are found in the open ocean, estuaries, and bays as well as inland freshwater river and lake systems; (c) they are subject to fishing and habitat impacts through all the various ecosystems; and (d) they experience a long time to maturity (8 to 24 years). Management measures in the commercial fishery include a coast-wide overall quota for yellow eels; the prohibition on catching silver eels except in the New York portion of the Delaware River with limited entry; no glass eel fishery in New York; a minimum size; and a minimum mesh size for eel pots. Management measures in the recreational fishery include a minimum size and a possession limit. There is no FMP for eels for waters outside 3 mi. of the coast.

FISHES COMMONLY SEEN IN EELGRASS BEDS

American eels and other fishes utilize eelgrass beds, which provide a variety of ecosystem services, for example, functioning as a sediment trap and removing excess nutrients from the water column. For marine invertebrates and juvenile fishes, eelgrass is a critical habitat serving as a nursery and providing a source of food and a haven to avoid predators.

In Long Island bays, the greatest diversity of fishes can be found during early fall in the vicinity of eelgrass beds. Over the course of four decades, under those circumstances, in total 85 different species were collected in Shinnecock Bay. In addition to the many temperate zone resident species, some of those fishes were juvenile warm-water southern species that entered via an inlet from the ocean during the summer on the flood tide. At that time those "tropicals" tend to disperse to the eelgrass margins, where they take refuge in the eelgrass meadow and become concentrated there, unable to leave safely and find their way back to the open ocean. In general, the number and diversity of fishes in the bay was greater during those years when eelgrass beds were extensive and healthy. Before harmful algal blooms destroyed some of the eelgrass beds in Long Island bays, eelgrass in those areas was dense enough to impede small boat operation.

The following is a list of eel grass community fishes that would only be found most predictably in early fall. Many of the southern species that would be found only infrequently are not included.

American eel, bay anchovy, Atlantic tomcod, oyster toadfish, Atlantic silverside, sheepshead minnow, mummichog, striped killifish, fourspine stickleback, threespine stickleback, lined seahorse, northern pipefish, northern searobin, grubby, black sea bass, bluefish, scup, weakfish, spotfin butterflyfish, tautog, cunner, naked goby, summer flounder, windowpane, winter flounder, hogchoker, planehead filefish, and northern puffer.

CONGER EEL
Family Congridae (*congrios*)

This is a large family of eels with 32 genera and 160 species found worldwide from temperate to tropical marine waters. The family includes species' sizes ranging from the local large conger eel to the slim-bodied garden eels found in warmer waters. As in all members of the eel order (Anguilliformes) formerly known as the Apodes, none of the eels have pelvic fins and they all have a large number of vertebrae in their vertebral column.

Conger eel

Conger oceanicus

Other common names: American conger
Gulf of Maine to eastern Gulf of Mexico
Max. length = 6 ft., weight ≥ 20 lb.

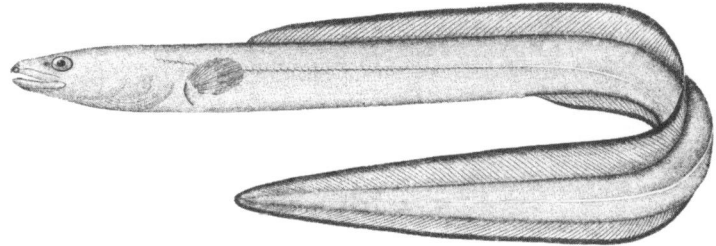

FIELD CHARACTERISTICS. The American eel and the conger eel have many distinctions other than maximum size. In the conger eel, the dorsal fin originates close behind the tip of the pectoral fins and the upper jaw projects beyond the lower jaw.

ECOLOGY/LIFE HISTORY. Although the geographic range is broad, like most fishes, conger eels tend to be more common in the middle of that range. They can be found nearshore in rocky coastlines but also offshore and sometimes inhabiting tilefish burrows and depths exceeding 300 ft.

A BRIEF HISTORY OF *LEPTOCEPHALUS*

In the mid-nineteenth century, a number of fishes that were small, elongate, transparent, and flattened with a minute head and small mouth were classified as being members of the genus *Leptocephalus*. That term referred to the slender head common to all. In 1864, T. N. Gill, an ichthyologist at the Smithsonian Institution, suggested that at least one of these species was actually a larval conger eel. It is now known that this larval form is common to the 15 different families of eels. Curiously, in the late nineteenth century's definitive descriptive catalogue of the fishes of North and Middle America, David Jordan and Barton Evermann still placed the conger eels into the family Leptocephalidae rather than Congridae, and they still had the conger eel in the genus *Leptocephalus* rather than *Conger*. Jordan and Evermann knew that the term *leptocephalus* was used erroneously to apply to the larval forms of different adult eel species and they knew that "leptocephalus" should not be used as the generic name, but they were required by the strict laws of priority to retain that genus. Thus, these early scientists retained the generic term and identified local conger eel as *Leptocephalus conger*. Subsequent work by other taxonomists led to the current scientific name.

ANCHOVIES
Family Engraulidae (*anchoas*)

This is a large family (16 genera with 139 species) that is primarily marine and is found in the coastal waters of the Atlantic, Indian, and Pacific Oceans. At present, these schooling fishes are harvested for some human consumption but more commonly the catch is sold as bait or processed into fish meal. Anchovy fisheries were famous for their abundance. However, in the past, there were some instances, as in the case of the Peruvian anchoveta, overfishing beyond their maximum sustainable yield reduced anchovy numbers. This decrease was further aggravated by a climate event (El Niño) along the Peruvian coastline that depressed the upwelling of nutrients, leading to a depletion of the anchovies' plankton food supply. At present, that fishery has fully recovered and is one of the largest fisheries in the world. Formerly, the Pacific coast northern anchovy, *Engraulus mordax*, was a source of commercial anchovies. When that fishery crashed, the source for edible anchovies in the United States ultimately became the European anchovy, *E. encrasteolus*, harvested from the Mediterranean.

Anchovies and herring are closely related families. Both of these families possess abdominal pelvic fins, pectoral fins placed low on the body, a forked caudal fin, and a single dorsal fin located midbody. All the fins are soft rayed. These two families possess smooth-edged cycloid scales characteristic of lower bony fishes but differ in two conspicuous ways. Herrings generally have sharp abdominal scutes running along the ventral edge of their body. Anchovies do not. Further, anchovies uniquely have very long upper jaw bones (maxillaries) that reach well beyond the eyes. Herrings have mouths that do not extend that far.

Of the 11 anchovies that occur in the Atlantic, only 2, the striped anchovy and the bay anchovy are found in New York waters.

Striped anchovy

Anchoa hepsetus (anchoa legitima)

Nova Scotia to Florida, Gulf of Mexico, Brazil (more common Chesapeake Bay to Florida)

Max. length = 6 in.

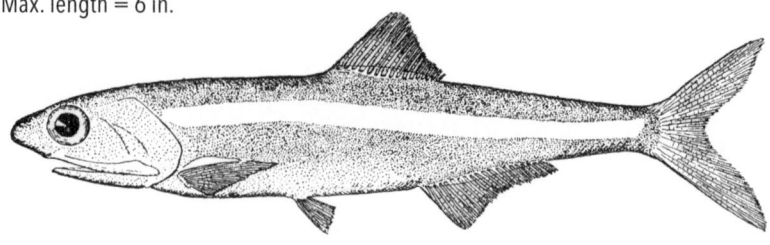

FIELD CHARACTERISTICS. The broad and distinct silvery stripe extends from the gill cover to the base of the caudal fin. The width of the stripe is equal to the eye diameter. The anal fin origin is under the mid or posterior part of the dorsal fin. In contrast, the silver stripe of the bay anchovy is less distinct and only as wide as the eye pupil diameter. Further, the anal fin origin is underneath the anterior margin of the dorsal fin.

Bay anchovy

Anchoa mitchilli (anchoa de caleta)

Maine to the Gulf of Mexico
Max. length = 4 in.

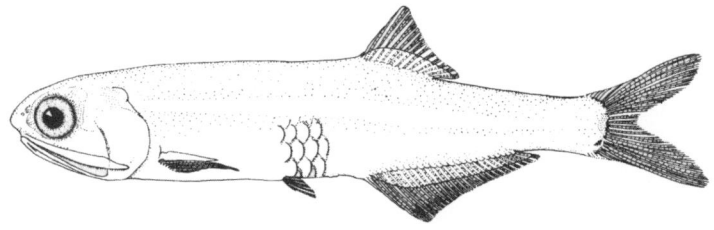

FIELD CHARACTERISTICS. The bay anchovy is a relatively common estuarine species in New York, but it is even more abundant along the Gulf Coast.

ECOLOGY/LIFE HISTORY. Bay anchovies are selective particle feeders, primarily eating zooplankton. Among the zooplankton, copepods are commonly found in the bay anchovy's diet although adult anchovies also feed on fish and crab larvae. Bay anchovies serve as common prey species for bluefish, striped bass, summer flounder, weakfish, sea birds, and marine mammals. Thus, the bay anchovy serves as a valuable member of a system that transfers energy from lower to higher trophic levels.

The species is named after Samuel Latham Mitchill, who in 1815 was the first naturalist to catalog the known fishes of New York. It was the zoologists, George Cuvier and Achille Valenciennes, who honored Mitchill in 1848 by naming this anchovy, *Engraulis mitchilli*. After a series of changes to the genus name, the ichthyologist, Samuel Hildebrand in 1943 finally established the name *Anchoa mitchilli* that we use today.

HERRINGS
Family Clupeidae (*sardinas*)

This family includes many of the fishes commonly known as different varieties of herrings, menhaden, sardines, and shad. Most of the 188 species of clupeids swim in large schools and are phytoplankton and/or zooplankton feeders. As such they have significant commercial and ecological value throughout their worldwide, predominantly marine, distribution. The herring family is characterized by having a single soft-rayed dorsal fin and abdominal scutes that, with varying degrees of serration, create a sharp ventral edge to their belly. Further, their compressed, silvery bodies are covered by scales that easily detach when the fish is handled.

Blueback herring

Alosa aestivalis

Other common names: river herring, glut herring, blackbelly
Nova Scotia to northern Florida
Av. adult = 10 in.

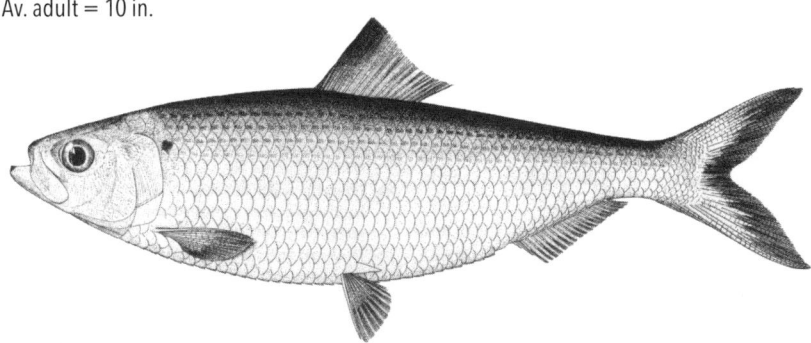

FIELD CHARACTERISTICS. Morphologically, the two river herrings (blueback herring and alewife) are very similar. Each is about the same size, females tend to be larger than males, and they both have a dark spot just behind their gill covers. Some differences include the dark rather than pale abdominal cavity lining in the blueback herring and, more apparently, the diameter of the eye in the alewife is greater than the snout length. Blueback herring range farther south than alewives. These "river herrings" and shad are anadromous. The term "anadromous" is derived from the Greek "up running," describing the upstream migratory movement of spawning fishes. *Alosa aestivalis* also tends to move into larger streams to spawn and does that later in the year than the alewife. Appropriately, the species name *aestivalis* means "of the summer" in Latin.

Hickory shad

Alosa mediocris

Other common names: shad herring, fall herring
Maine to Florida
Max. length = 18 in.

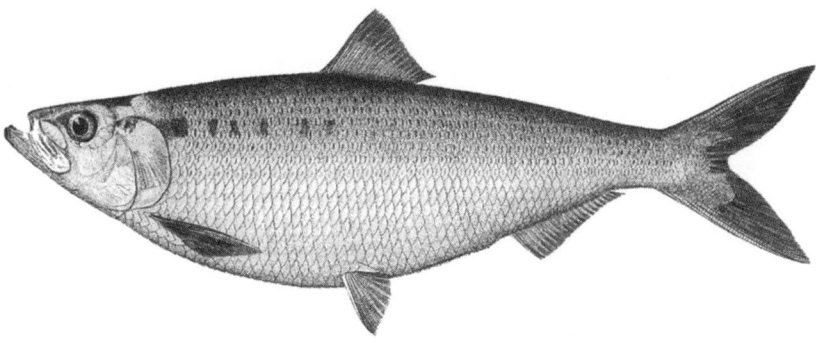

FIELD CHARACTERISTICS. The tip of the lower jaw extends noticeably beyond the upper jaw when the mouth is closed, in contrast to the lower jaw position in other *Alosa* species (i.e., American shad, alewife, and blueback herring). The species name *mediocris* might suggest that at the time, in the early nineteenth century when this fish was named, its bony condition and flesh was not highly valued as food. This is still the case although the roe is prized.

Alewife

Alosa pseudoharengus

Other common names: river herring, glut herring, sawbelly
Labrador to South Carolina
Av. adult = 10 in.

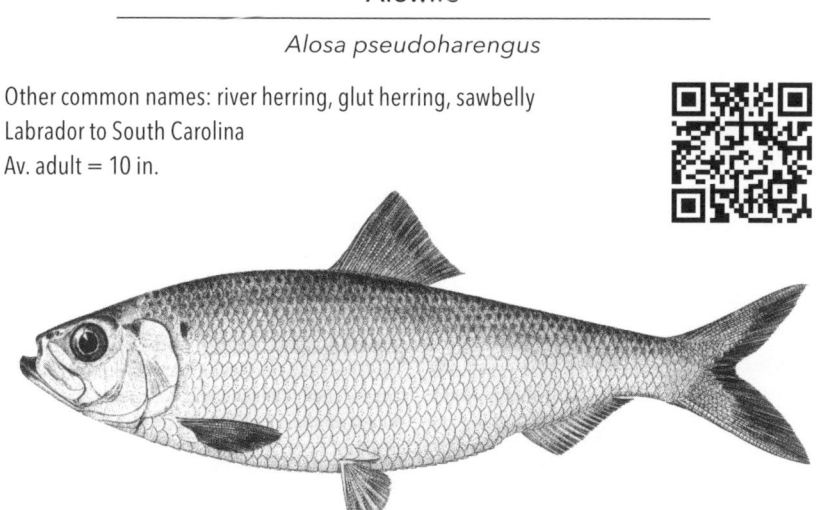

FIELD CHARACTERISTICS. Most alewives are best seen as they migrate upstream. Their dorsal area is a grayish green, but the sides and ventral area are silvery. Their silvery remains are revealed when at times they fail to survive the rigors of migration and their bodies are seen along the shoreline after being scavenged by gulls, egrets, herons, and raccoons.

ECOLOGY/LIFE HISTORY. Beginning in March, alewives move upstream, and when in quiet shallow waters, a gravid female will tend to attract many males and during circular movements both broadcast their gametes. A large female may have well over 100,000 eggs to be fertilized. Alewife populations occur from Canada to the Carolinas and thus are presented with a variety of ecologies that likely influence aspects of their spawning behavior. For example, adults who survive can be repeat spawners, that is, return to spawn the following years. Further, some may spawn more than once in a season, and other individuals are believed to spawn in their natal streams. Using their keen olfactory sense, alewives have been shown to prefer water from their parent pond suggesting that this mechanism may be used to return to a natal stream.

Although there are some minor commercial uses of alewives, their broad ecological value is more significant. These anadromous fish transport valuable nutrients from the ocean to freshwater, and those nutrients are released in the form of dying post-spawners or as prey to a variety of bird and mammalian predators as the spawning alewives moved upstream. In their ocean and bay phases, alewives are valuable food sources for large piscivores such as striped bass, bluefish, weakfish, ospreys, and seals.

Finally, although there are no land-locked alewife populations on Long Island, there are such populations in many of New York's freshwater lakes including the Finger Lakes and Great Lakes.

LITERATURE. Thunberg 1971; Kissil 1974; Loesch 1987; Jessop 1994; Helfield and Naiman 2001; McCartin et al. 2019

FISHERIES. As with many other inshore easily caught species, commercial catches of alewives were greatest in New York from the late 1800s through the 1930s except for the anomaly in 1966, when four million pounds were caught to make fish meal in New York's last reduction factory (a reduction factory was a facility that rendered various fishes, but mostly clupeids, into fish meal and oil). Except for this 1966 anomaly, landings peaked in the late 1890s at around 2.5 million pounds. Landings were generally 50,000 lb. or less during the second half of the twentieth century and into the twenty-first century. Alewives have historically been harvested with dip nets, gillnets, and pound nets. Catches primarily occurred in New York's inshore bays and river systems including Long Island small rivers and the Hudson River.

Alewives are not a very popular recreational fish and catch reporting is very sporadic. They are not caught recreationally with hook and line but with dip nets and cast nets and are often used as bait to catch other species.

MANAGEMENT. Along with blueback herring, alewives are managed as river herring by the ASMFC and the NYSDEC through the FMP for shad and river herring within 3 mi. of the coast and inland waters (ASMFC 1985a). There are numerous stocks of river herring along the coast (based on river systems), and most are in a depleted condition, including New York stocks (ASMFC 2017a). The primary causes of the depleted condition of river herring stocks are due to many factors, including directed and incidental fishing, habitat loss or other habitat impacts (including the historical damming of streams and rivers), and predation. Some of these problems are being addressed by local communities and organizations committed to conserving anadromous fish runs.

Alewives, like other anadromous fishes, have freshwater lifecycle requirements and thus must come into close contact with human populations and are thus vulnerable to the many threats and potential sources of mortality associated with activity in and around river and stream systems. These issues

NEW YORK'S ANADROMOUS SPECIES

Petromyzon marinus
  sea lamprey
Acipenser oxyrhincus
  common Atlantic sturgeon
Acipenser brevirostrum
  shortnose sturgeon
Anguilla rostrata
  American eel (catadromous)
Alosa aestivalis
  blueback herring
Alosa mediocris
  hickory shad
Alosa pseudoharengus
  alewife
Alosa sapidissima
  American shad
Osmerus mordax
  rainbow smelt

Oncorhynchus mykiss
  rainbow trout
Salmo salar
  Atlantic salmon
Salmo trutta
  brown trout
Salvelinus fontinalis
  brook trout
Microgadus tomcod
  Atlantic tomcod
Gasterosteus aculeatus
  threespine stickleback
Morone americana
  white perch
Morone saxatilis
  striped bass

present special problems to the assessment and management of alewives. To help alewife populations recover, current regulations in New York prohibit the commercial and recreational catch of alewives except in the Hudson River, where there is a low possession limit. There are also habitat improvement activities taking place to help with alewife recovery.

There is no federal FMP for alewives for waters outside 3 mi. off the coast. However, river herring bycatch is addressed through a river herring bycatch cap in the mackerel and Atlantic herring fisheries. There is also a fishing industry proactive bycatch avoidance network coordinated by the Cornell University Cooperative Extension Marine Program. The avoidance network uses real-time fishing reports to help fishermen avoid concentrations of shad and river herring and thus reduce bycatch interactions.

American shad

Alosa sapidissima (sabalo Americano)

Nova Scotia to Florida
Av. adult = 19 in.

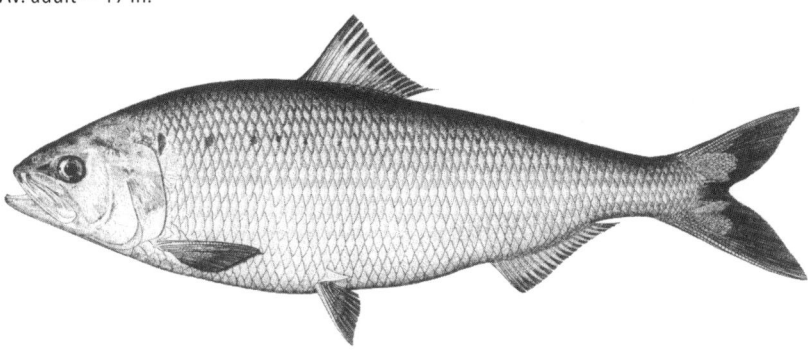

FIELD CHARACTERISTICS. This is the largest member of the herring family. The reported maximum length is 30 in., as compared with the approximate maximum lengths for the Atlantic herring (15 in.), alewife and blueback herring (16 in.), and Atlantic menhaden (20 in.).

Herring family members have distinctive qualities that aid in their identification. For example, the American shad has a relatively small head that is equal to one-fourth the shad's standard length as compared with menhaden's larger head, which is equal to one-third the fish's standard length. (Standard length is the distance between the snout and the base of the caudal fin). Further, the shad's pelvic fin has eight rays whereas the menhaden has seven pelvic rays, and the body scales on the menhaden are serrated whereas the scales on other family members are rounded.

Like the other *Alosa* species commonly found in the waters around New York, the shad has a large, dark shoulder spot located behind the operculum (gill cover). The spot is not as well defined as the classic ocellus (eye spot) found in a variety of other fishes. Thus, the function of *Alosa*'s shoulder spot may not serve to deceive predators regarding the location of the true eye and head. The large spot on the American shad is often followed by a single row of several smaller, sometimes indistinct, spots. The menhaden's shoulder spot is often followed by many irregular smaller spots arranged along the trunk in multiple rows. The alewife has only one large shoulder spot. An additional distinction between the American shad and the alewife is that in the shad, the upper jaw extends to a vertical line through the rear margin of the eye whereas in the alewife, the upper jaw reaches only to a vertical line through the center of the eye.

ECOLOGY/LIFE HISTORY. The American shad is native to the East Coast, but in the 1870s this shad and the striped bass were independently introduced into Pacific waters for commercial and recreational purposes, although now they are primarily sport fishes. Both of these species are very edible. In the case of the shad, the flesh is fine although as in many herring-like fishes it is bony. The name *sapidissima* is derived from the Latin word meaning "most delicious." Of particular food value to fishermen is the shad roe collected from females during their spring spawning migration.

Some American shad migrate significant distances along the Atlantic coast during their spawning migration. For example, after overwintering in northern Florida they may move as much as 2000 mi. north in the spring following their preferred temperature range of 55 to 64 °F (13–18 °C). There are spawning populations of shad from Florida to Canada. It has been reported that all the shad in Florida populations die after spawning, but 55%–77% of New Brunswick spawners survive and are potentially available to spawn in the following year. Curiously, the fecundity (a measure of fertility) of New Brunswick shad is lower than the fecundity of more southern populations. It is suggested that northern rivers are harsher environments for eggs and larvae and therefore those northern fish invest their energy stores into improving post-spawning survival and increase the likelihood of returning to spawn in the following year. Other investigators have provided good evidence that American shad return to their natal stream after spending up to 4 years at sea. While in the ocean, shad feed on zooplankton, but like many adult anadromous fishes, they do not feed during their spawning run.

The shadbush is a North American bush or tree that was named by early colonists. The plant is one of the first native plants to bloom (white blossoms) in the spring, coinciding with the time of the year (early May) when shad migrate upstream. The spring appearance of lilacs is a commonly recognized signal that our most popular game fishes (e.g., stripers and tautog) have returned.

LITERATURE. Leggett and Carscadden 1978

FISHERIES. As with many other inshore easily caught species, commercial catches of shad were greater during historical times than they have been from the mid-twentieth century on. The highest recorded catches of shad in New York occurred in the 1880s and 1890s, and there are accounts from colonial times of large shad catches. In fact, shad were a staple for both colonists and Indigenous Americans, particularly along the Hudson River. Shad were harvested both for their flavorful flesh as well as their very rich and tasty roe. Shad were similarly important to coastal communities along the entire Atlantic coast, and catches of shad once supported the largest and most important food fisheries along the Atlantic coast before World War II. In-river landings began decreasing and ocean harvest landings began increasing during the 1970s as that fishery developed.

Shad has been a popular recreational fishery, particularly in the Hudson River. However, the size of recreational catches is largely unknown. The Marine Recreational Information Program estimates the number caught by recreational anglers, but estimates of shad are imprecise and are not useful for management purposes as this survey does not capture data on recreational fisheries in inland waters like commercial surveys do.

ALL TACKLE WORLD RECORD: 11 lb. 4 oz., Connecticut River, Massachusetts 1986

NEW YORK RECORD: 9 lb. 4 oz., Hudson River, 2007

MANAGEMENT. Shad is managed by the ASMFC and NYSDEC through the Interstate FMP for Shad and River Herring (ASMFC 1985a). The most recent shad stock assessment (ASMFC 2020) determined that shad stocks were at all-time lows and did not appear to be recovering to acceptable levels. The primary causes for continued stock declines were identified as a combination of excessive total mortality (natural, fishing, and other human induced), habitat loss and degradation, as well as migration and habitat impediments. Shad stocks are defined by each river system that historically produced shad.

Anadromous fish species, like shad, are unlike most other marine species that spend their lives in marine and estuarine environments. Anadromous fish as a result of their freshwater life cycle requirements must come into close contact with human populations and are thus vulnerable to the many threats and potential sources of mortality associated with human activity in and around river systems. These issues present special problems to the assessment and management of shad. Current regulations in New York prohibit the commercial and recreational catch of shad, except in the Delaware River basin.

Atlantic menhaden

Brevoortia tyrannus

Other common names: bunker, mossbunker, fatback, pogy
Nova Scotia to Florida
Av. and max. length = 12 and 18 in.

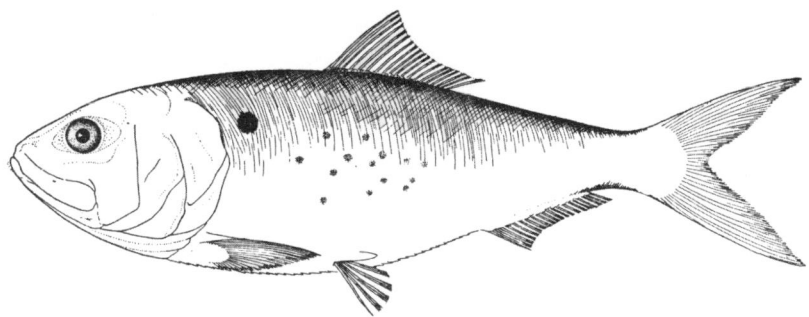

FIELD CHARACTERISTICS. Menhaden have large heads, deep bodies, and a
conspicuous dark shoulder spot followed by a variable array of smaller dark
spots on the trunk. The scales are cycloid but have fine comb-like edges at the
rear of each scale.

ECOLOGY/LIFE HISTORY. Menhaden are toothless but have hundreds of long
gill rakers on their gill arches that assist filtering plankton from the water
that enters through their wide-gaping mouth. Curiously, the gill rakers do
not simply act as dead-end filters and sieve out the plankton. As in another
herring family member (American shad), 95% of the suspended food particles
do not come into contact with the gill rakers but instead remain suspended
as the water flows parallel to the filter surface and leaves between the rakers.
However, in menhaden the particles become more concentrated as they travel
toward the esophagus after which the fish swallows the concentrated slurry
of food.

   Menhaden are effective filter feeders. They swim continuously, and
an adult moving in a school may filter about 350 gallons of water per
hour. Considering that some large schools may contain 100,000–200,000
menhaden, it is possible, in some cases, to see plankton-rich water being
cleared behind the school as the school swims forward. The plankton
being filtered range from 16 µm diatoms (phytoplankton) to relatively large
zooplankton, for example, 1200 µm copepods and 10,000 µm (10 mm)
shrimp. Menhaden also may have significant negative effect on populations of
crustaceans and fishes by feeding on their small larval forms.

   Adults do not occur beyond the continental shelf. In the simplest terms, menhaden move south in the fall and north in the spring, at which time they spawn. The resulting young move into estuaries that act as nurseries where there are abundant supplies of food for them.

At times when menhaden schools are close to shore, a school of large and actively feeding bluefish can force large numbers of menhaden to concentrate in shallow waters, causing hypoxic conditions and the mass death of a menhaden school. Menhaden are sensitive to low oxygen levels. Water contains only about 3% of the oxygen that can be found in the same volume of air. So oxygen is easily depleted by a large school of confined and active fish. As a defense, menhaden have a gill area, across which oxygen can diffuse, that is more than 18 times the area of the body surface excluding fins.

Menhaden are a valuable prey object for all the marine carnivores in our region. Those predators include striped bass, bluefish, weakfish, summer flounder, whiting, Atlantic cod, sharks, tuna, marine mammals, and ospreys. Menhaden is one of the preferred baits for striped bass fishermen and is the largest American commercial fishery. The catch is processed for animal feed, fertilizer, and oil. The latter is the base for health food supplements. A reduction in large schools of filter feeding menhaden may also impede their ability to assist in reducing harmful algal blooms.

The species name *tyrannus* means ruler in Greek, the name given this species by a nineteenth-century ichthyologist. Apparently, even at that time, this fish played a significant role in the economy and ecology of the marine environment.

LITERATURE. Brainerd 2001; Sanderson et al. 2001

FISHERIES. Menhaden has been harvested since colonial times, originally for use as fertilizer. Although it is edible, its high oil content and many bones have discouraged its use as human food. Harvest has been historically for reduction into fish meal and oil or for bait. As with many other nearshore species in New York, menhaden experienced its highest landings in the 1800s and early 1900s. Peak landings of over 288 million pounds occurred in 1880. Menhaden is a plentiful species, and their use for reduction and bait results in a high volume but low value per pound. In the 1800s and early 1900s there were several menhaden reduction factories in New York that rendered menhaden. These were primarily located on the east end of Long Island. The menhaden reduction facilities went through a consolidation in the 1940s through 1960s as many of the reduction plants closed or were bought out and production moved to fewer, larger facilities in New Jersey, Virginia, and other southern states. There was also much community opposition to the odorous operation of the reduction facilities as the Long Island population centers expanded into previously remote areas where the facilities were located. The last remaining reduction plant in New York, located on Napeague, the area between Amagansett and Montauk, closed in the mid-1960s and moved its operation to New Jersey. Owing to its high-volume fishing, menhaden dominated New York landings in the 1800s through the mid-1960s, and most years the volume of menhaden landings were greater than most, if not all, other species combined. Landings during the 1940s through the 1960s averaged over 90 million pounds per year. Until the closure of the last reduction facility, purse seines were the primary gear used to catch menhaden in New York. Since 2013 New York menhaden landings have been around 1 to 2 million pounds per year with a recent high of 4.4 million pounds in 2020. Menhaden in New York are currently caught with beach seines, pound nets, and gillnets. It is currently illegal to catch menhaden with purse seines in New York waters. New York menhaden landings are used as bait for lobster pots, crab pots, and recreational fishing. There is no recreational fishery for menhaden.

MANAGEMENT. Menhaden is sustainably managed by the ASMFC and NYSDEC through the Atlantic Menhaden FMP (ASMFC 1981a). Menhaden is currently at a high level of abundance and is not overfished nor experiencing overfishing (ASMFC 2022). The fishery is managed by an overall commercial quota that is then divided among the states based on landings in 2018, 2019 and 2021. Virginia has the largest allocation because of the reduction fishery that still occurs there. The menhaden FMP is unique among Atlantic coast FMPs. In setting the overall quota, the FMP uses ecological reference points that consider menhaden's role as a major forage fish for predator species such as striped bass. In addition to the state quota, New York also imposes trip limits to regulate the commercial fisheries. Since there is no recreational fishery for menhaden, there are no recreational management measures.

Atlantic herring

Clupea harengus

Other common names: sardine herring, sea herring
Greenland to North Carolina and eastern Atlantic Ocean
Av. and max. length = 9 and 17 in.

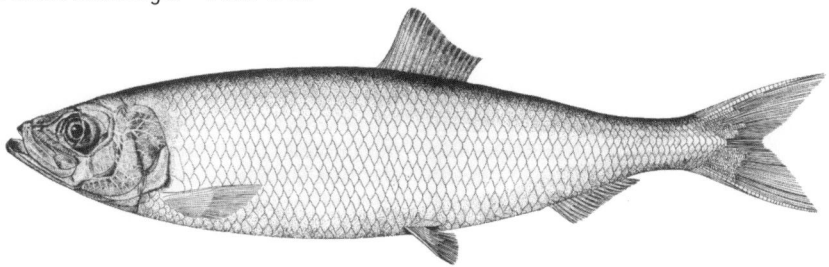

FIELD CHARACTERISTICS. Unlike menhaden, shad, and alewives, there is neither a prominent dark spot immediately behind the gill covers nor does the tip of the lower jaw fit into a median notch in the upper jaw. Herring have small teeth on their lower jaw and vomer (front median surface of the mouth's roof). Small jaw teeth can also be found in the shad and alewife, but there are no vomerine teeth.

The belly is rounded and the ventral scutes are not sharply keeled although the scales meet to form a weak ridge. This ridge is in contrast with the other common clupeids that have serrated ventral scutes resulting in a sharp keel.

Silvery platelets on the surface of the scales are capable of reflecting light in such a way as to act to obscure the schools of herring from their many aquatic predators. The scales, as in many clupeids, are in shallow pockets and are easily rubbed off when the fish is handled. These deciduous scales are replaceable.

ECOLOGY/LIFE HISTORY. Migration patterns generally follow the seasonal development of zooplankton. This herring migrates seasonally within open ocean basins (oceanodromy). Their movement is often cyclonic in that adults will move against oceanic currents to the spawning area where their small pelagic larvae might find rich, appropriately sized food supplies, after which the young herring drift "downstream" back to the adult feeding area. In the northeastern Atlantic, different spawning groups (stocks) occur in separate areas. These different stocks may spawn at different times from spring to fall. These stocks may also differ in number of fin rays, vertebrae, scales, and gill rakers.

Herring have a busy vertical migration pattern. During the day herring tend to be in deeper water, then migrate up at sunset; they migrate back down at midnight, up again near dawn, and downward as the sun rises—all of which is the result of following their food.

LITERATURE. McKeown 1984; Lindgren et al. 2011

FISHERIES. Unlike many states to our north, Atlantic herring has not been an important commercial fishery in New York. For the past 50 years, New York commercial landings have been generally less than 100,000 lb. and often less than 50,000 lb. Historic landings were also low (except for the anomaly in 1966 when 6.4 million pounds were caught to make fish meal in New York's last reduction factory). They are used as bait by recreational fishermen and as lobster pot bait. There is no recreational fishery in New York for Atlantic herring. The fishery is so small in New York compared with states to our north, that New York has de minimus status (only a minor role due to low landings) in the management of Atlantic herring.

Gizzard shad

Dorosoma cepedianum (sardina molleja)

Quebec to Florida, Gulf of Mexico and the estuarine waters along the Atlantic and
   Gulf coasts
Max. length = 20 in.

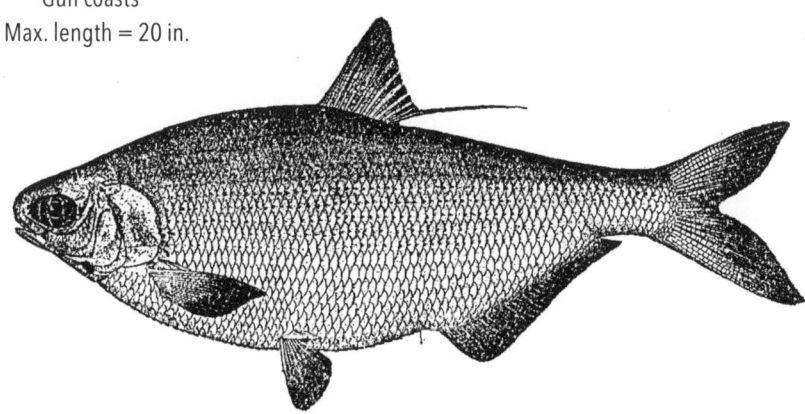

FIELD CHARACTERISTICS. The gizzard shad has a compressed body and a round
snout projecting over a mouth that seems to open somewhat ventrally. More
conspicuously and like some other more common clupeids, the gizzard shad
has a large spot behind the gills. However, in this species the last ray of the
dorsal fin is elongated, which is a feature it shares with only the Atlantic
thread herring, a species rare enough in New York waters so as to be included
only within a list of species not predictably encountered.

Round herring

Etrumeus teres (sardina japonesa)

Maine to Florida, Gulf of Mexico, Venezuela
Max. length = 9.8 in.

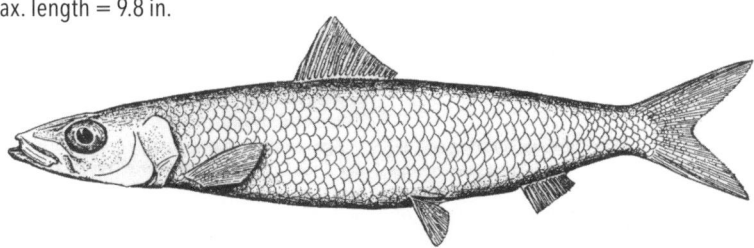

FIELD CHARACTERISTICS. The round herring is seen more commonly offshore
than in the bays. As the common name suggests, the belly is rounded and
does not have the saw-like keel that typifies many herring family members.

SMELTS
Family Osmeridae (*capellanes*)

The 31 members of the smelt family are distributed in the Arctic and northern latitudes of the Atlantic and Pacific Oceans. They are primarily marine but have anadromous and freshwater representatives. Most of these species are obscure, although in Canadian waters capelin is an important commercial species in the fish and oil industry. Most family members are relatively small and occur in dense schools and as such are an important food source for piscivores.

This family has the array of anatomical features that characterizes ancestral teleostean fishes. Those features include cycloid rather than ctenoid scales and an adipose fin. The adipose fin is a small, fleshy, rayless fin on the rear dorsal part of the trunk. This latter fin is found in a variety of fish families such as salmon, catfishes, characins, and other families that tend to represent fishes that are considered to be more primitive in the evolutionary history of teleostean fishes. Those early teleosts tend to have a set of characteristics in common, including a single soft-rayed dorsal and anal fin. The function of the adipose fin is uncertain, although because it is placed above the single anal fin, the adipose fin could serve as a stabilizing counter balance.

Rainbow smelt

Osmerus mordax

Labrador to Delaware River
Max. length = 14 in., av. adult = 8 in.

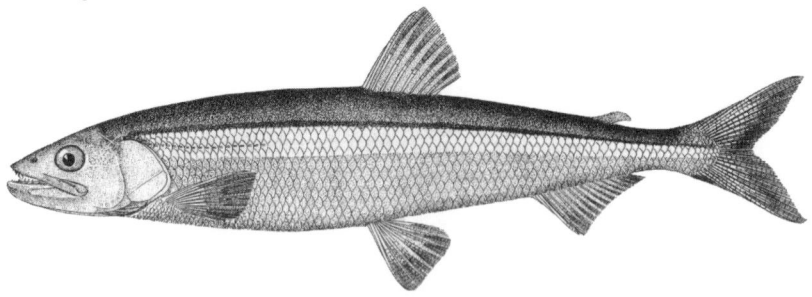

FIELD CHARACTERISTICS. The smelt has a pale body with a silvery stripe running along its straight lateral line to the base of its deeply forked caudal fin but because of a variety of differences, especially the teeth, this fish would not be mistaken for Atlantic silverside.

ECOLOGY/LIFE HISTORY. The rainbow smelt is more common in Canada and New England. Massachusetts seems to be the southern limit to the presence of large populations. It is not common on Long Island and is, like other anadromous species, threatened by obstructions in the spawning streams. The smelt, alewife, blueback herring, and the glass eel stage of the American eel may use the same stream during early spring. Migrating smelt tend to be the earliest migrators and do not travel far upstream. When they occur in significant numbers, smelt are caught by recreational fishermen using rod and reel or, more traditionally, by dip net as they migrate upstream. Smelt that are in landlocked lakes are ice fished during the winter. Regardless of how caught, fried smelt are considered a delicacy.

This genus *Osmerus* (Greek for odorous) suggests that in this case, a fresh fish has a pleasant odor. The species *mordax* (Greek for biting) is more noteworthy because the rainbow smelt is very carnivorous. As such, it has a highly developed set of teeth. These teeth are not simply on the upper and low jaws but include fang-like teeth on the tongue and teeth on the roof and sides of the mouth. The smelt feeds on crustaceans, worms, squid, and small fishes, and its predators include cod, salmon, bluefish, striped bass, and seals.

LIZARDFISHES
Family Synodontidae (*chiles*)

This benthic family is found in the warm marine waters along the Atlantic, Pacific, and Indian Ocean coastlines and includes four genera and 57 species although only 1 species would be encountered locally. In addition to the enlarged pelvic fins, members of the lizardfish family have an adipose fin.

Inshore lizardfish

Synodus foetens (chile apestoso)

Other common names: galliwasp, soapfish
Massachusetts to Brazil
Max. length = 16 in.

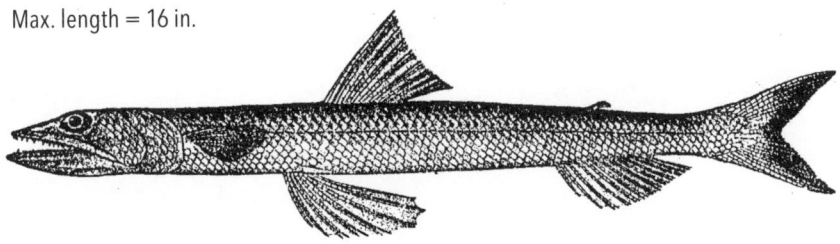

FIELD CHARACTERISTICS. The lizardfish is aptly named because it acts as a lie-in-wait predator and has an elongate, lizard-like, cylindrical body. Certain features make this species an effective predator; for example, it has a large mouth and a gape that extends well behind the eyes. Further, there are sharp depressible teeth in the upper jaw and other teeth on the tongue and lower pharyngeal (throat) bones that assist in grasping and holding on to the prey.

ECOLOGY/LIFE HISTORY. Within its range, it is most common in warmer waters from shallow sand flats, among sea grasses, adjacent to coral reefs, or at depths nearing 300 ft. This fish's ability to ambush prey is facilitated by being able to rest motionlessly on their large pelvic fins and by having a body color and pattern that varies with the background.

MERLUCCIID HAKES
Family Merlucciidae

Not all "hakes" belong to the same family. The phycid hakes include such species as the red and spotted hakes who like the merlucciid hakes are closely related to the cod family. In both hake families, unlike the cods, there are only two dorsal fins, with the second being considerably longer. Species of the merlucciid family have no chin barbel and their pelvic fins are not the elongated "feelers" as seen in phycid hakes.

There are 22 species in this family. Hakes occur offshore and inshore on both sides of the Atlantic Ocean as well as worldwide in cool marine waters.

Silver hake

Merluccius bilinearis

Other common names: whiting, frostfish
Newfoundland to Florida
Max. length = 2.5 ft.

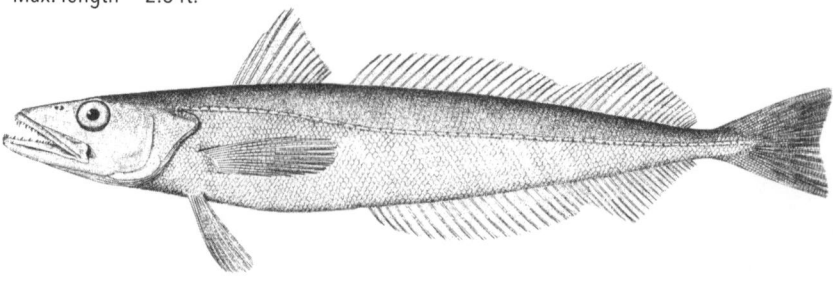

FIELD CHARACTERISTICS. The silver hake is indeed silvery. It has a pointed head and a projecting lower jaw. The genus *Merluccius* means "sea pike" probably because it, like pikes, are predatory and have large mouths with sharp teeth. The species name *bilinearis* refers to a lateral line that has the appearance of being double.

ECOLOGY/LIFE HISTORY. Adult silver hakes are primarily fish eaters and have the kind of profile associated with aggressive piscivores. Although it does not have the size and formidable tooth array as the solitary hunting northern pike or a barracuda, the silver hake is equally voracious and feeds on a variety of prey including young silver hake. During the winter months, silver hake come very close to shore to hunt the abundant fishes found in the surf zone. On occasion the hake's enthusiasm causes them to become stranded on the beach and freeze. Their alternative common name "frostfish" arises from that event.

FISHERIES. Silver hake has been a key component of New York's commercial fisheries since the 1930s and the development of the trawl fishery, often ranking in the top 5 or 10 by landed weight, of finfish species. They are usually marketed as whiting rather than silver hake. Their value, however, is typically modest, and they are considered a "cheap fish" because of their short shelf life, limited market demand at higher prices, and the fact that they do not freeze well. However, they have higher demand in Europe, and New York landings fluctuate depending on export opportunity and competition from Canada and other countries for the European market. Landings were very high in New York in the late 1980s through the 1990s due to strong export opportunities for silver hake. New York landings averaged around 10 million pounds annually during this period and peaked at 14.1 million pounds in 1998. Landings have since dropped off and since 2010 have averaged two to four million pounds annually. The fishery takes place primarily offshore with trawl gear. There has only been a very small recreational fishery in New York for silver hake as they are not a popular gamefish. Reported recreational catches are a couple thousand pounds annually at most.

ALL TACKLE WORLD RECORD: 4 lb. 8 oz, Maine

NEW YORK RECORD: None

MANAGEMENT. Silver hake is managed sustainably by the NEFMC and NMFS. There are two stocks of silver hake, a northern (Gulf of Maine and northern Georges Bank) and a southern (south Georges Bank through the Mid-Atlantic). Neither stock is overfished nor is overfishing taking place (NEFMC 2020). Silver hake is managed as part of the multispecies FMP (NEFMC 1985), which also manages traditional "groundfish" species such as Atlantic cod, haddock, and flounders. As such they are part of the "small-mesh multispecies" management unit and are managed under a mesh size exemption to allow the use of the small mesh needed to catch silver hake. The federal waters (outside 3 mi.) management measures incorporate gear and mesh size specifications, specific areas and seasons where the small mesh can be used, and overall quota and trip limits. There are no management measures in place for the recreational fishery. New York does not have any management measures for silver hake within New York waters.

PHYCID HAKES
Family Phycidae

This is primarily an Atlantic Ocean family. Four species occur in our region, but 21 other species are variously distributed along the North American coast, eastern Atlantic, Mediterranean and a few in the western Pacific and Indian Ocean. The family is characterized by possessing a long anal fin and second dorsal fin. They also have a chin barbel and long, filamentous, thoracic pelvic fins.

Fourbeard rockling

Enchelyopus cimbrius

Greenland to Gulf of Mexico and in the eastern Atlantic
Max. length = 16 in.

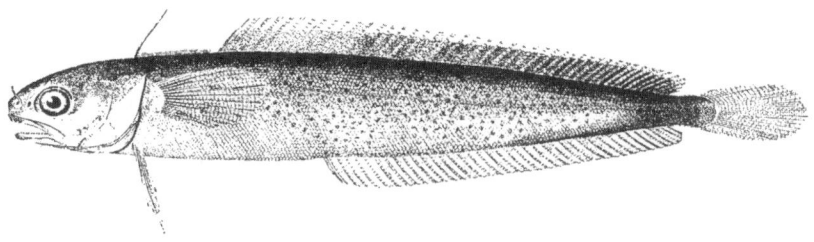

FIELD CHARACTERISTICS. Dorsal fin is preceded by one very long ray and a fringe of many short rays. The pelvic fins, in contrast to other phycid hakes, are short. In addition to the standard chin barbel, there are four barbels on the snout, thus the common name, "fourbeard." The elongated second dorsal has a dark blotch at the end. The anal fin has an elongate dark stripe, and the caudal fin has a dark blotch on its lower surface.

Red hake

Urophycis chuss

Other common names: squirrel hake, ling
Nova Scotia to North Carolina
Max. length = 19 in.

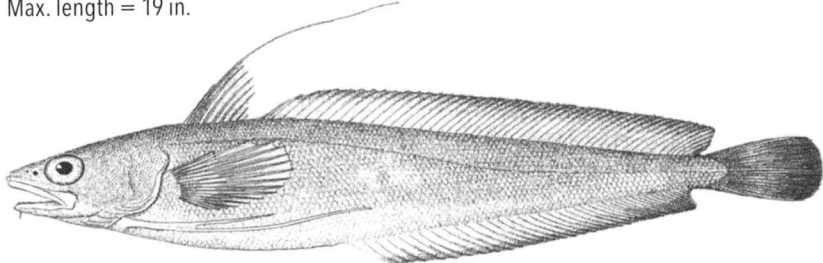

FIELD CHARACTERISTICS. The red hake has a lateral line scale number less than 118. The upper jaw may reach back to the rear edge of the pupil. Red hake tend to be found in deep water and as juveniles can be found within the mantle cavity of sea scallops or may find shelter underneath those shellfish. Other species, for example the inquiline snailfish that is occasionally encountered in our waters, also often lives in association with sea scallops.

Spotted hake

Urophycis regia (merluza barbona reina)

Massachusetts to Florida, Gulf of Mexico
Max. length = 16 in., common = 11 in.

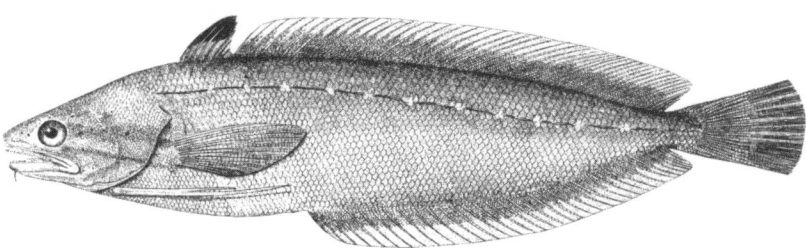

FIELD CHARACTERISTICS. Unlike other local family members, the spotted hake does not have any prolonged dorsal fin rays. The pectoral fins reach as far back as the anal fin, and the first dorsal fin has a distinct black blotch. The lateral line is marked along its entire length by an evenly spaced series of white spots.

White hake

Urophycis tenuis

Labrador to Florida
Max. length = 4.3 ft., common = 2.3 ft.

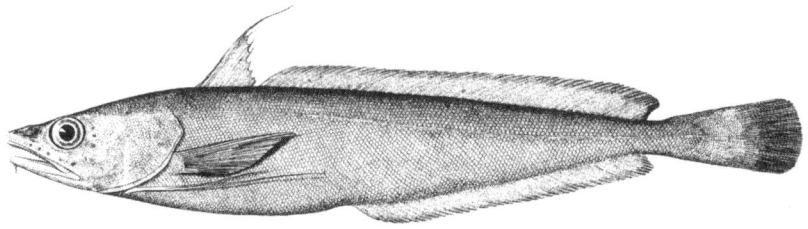

FIELD CHARACTERISTICS. Very similar to the red hake, both having the third ray in the first dorsal fin conspicuously prolonged. However, in the white hake, the lateral line scale number ranges from 119 to 148 and the upper jaw may reach back to the rear edge of the eye.

LITERATURE. Able and Musick 1976; Garman 1983

CODS
Family Gadidae (*bacalaos*)

Although there are a few exceptions, this is predominantly a cold-water marine family. It includes 31 species distributed within 16 genera. Representatives can be found at great depth, along the coast, and in brackish estuaries and rivers. The largest or most important commercial family members in or near New York are the Atlantic cod, pollock, and haddock. In general, the family is characterized by possessing three soft-rayed dorsal fins, two anal fins, and a short chin barbel (a sensory organ containing taste receptors).

Atlantic cod

Gadus morhua

Other common names: rock cod
Greenland to North Carolina and the eastern Atlantic Ocean
Max. size = 4.5 ft., 210 lb.

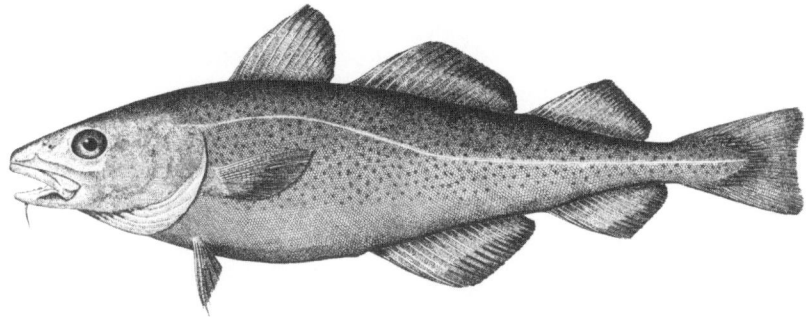

FIELD CHARACTERISTICS. The gadid family members are among the most important common fishes in our region and can be easily distinguished from each other. The Atlantic cod has a square or concave tail, a modest chin barbel (length = eye diameter), and pelvic fin rays that are not particularly elongate.

ECOLOGY/LIFE HISTORY. In general cod is a cold-water species. They are migratory and like many other fishes move in response to changing temperatures primarily for feeding and reproduction.

Spawning tends to be during the winter and spring although spawning can occur at other times depending on the location of this broadly distributed species.

Female cod can produce millions of eggs per year and these result in cods historically occurring in vast schools, making it possible to have hundreds of millions caught per year. When a cod fishery crashes it conceivably can disrupt predator–prey interactions affecting the entire North Atlantic ecosystem. Predators on large cod include sharks and seals.

Many of the cod's prey are benthic invertebrates, such as crabs, lobsters, and mollusks, but it also feeds on midwater squid and a variety of schooling fishes, for example, herring and sand lances. Barbels and pelvic fin rays in most cod family members are associated with taste buds and assist in detecting edible food objects. Curiously, some cod stomach content records also include normally inedible objects such as keys, other metal objects, and fabric. Like many other fishes, cods do not feed during spawning.

Cod does not tolerate warm waters and has an upper lethal temperature of about 68 °F (20 °C). Its ability to withstand near-freezing conditions is, like some other northern cold-water species, due to seasonally produced blood antifreeze.

Before North America was colonized, Portuguese fishermen were catching cod by hook and line off the coasts of Newfoundland and New England. In fact, after 1000 CE, fishing cultures shifted from freshwater to marine fishes such as the abundant cod and herring populations. The cod were particularly valuable because they were easily caught using hand lines, readily preserved by drying or salt curing, and were a rich protein source. Considering the many species of cod in the world, the Atlantic cod is the most likely species to be found on menus in New York and New England.

FISHERIES. Although an iconic fish and an important commercial and recreational fishery in New England, cod has been less important in New York, particularly during the past 20 years due to abundance issues, shifting distributions patterns, as well as changes in management that cut many New York commercial fishermen out of the codfish fishery. New York commercial landings of cod were highest during the late 1930s and early 1940s, when they ranged between 4 and 8.5 million pounds. Since the mid-1990s, commercial cod landings in New York have been less than 250,000 lb. and since 2012 less than 50,000 lb. Estimates of New York recreational cod landings are very imprecise and thus fluctuate wildly. Landings have recently fluctuated between 0 pounds and 1.7 million pounds, although they have mostly been in the 100,000 to 200,000 lb. range.

ALL TACKLE WORLD RECORD: 103 lb. 10 oz., Norway 2013

NEW YORK RECORD: 85.0 lb., 1984

MANAGEMENT. Cod is managed jointly by the NEFMC, NMFS, and the NYSDEC. The NEFMC and NMFS manage cod as part of the Northeast Multispecies (groundfish) FMP (NEFMC 1985), which also includes a variety of other species including haddock, windowpane, and winter and yellowtail flounder. The fishery is managed as two separate stocks: Georges Bank and Gulf of Maine. Cod from the Georges Bank stock are also found off New York and Rhode Island. The Georges Bank stock is overfished, but the overfishing status cannot be determined (NMFS, 2022a). Fishing is still allowed but at reduced levels. Both stocks of cod have been in a rebuilding mode for many years as part of the FMP. This rebuilding program is based on overall quotas, minimum size, seasons, gear restrictions, trip limits, and individual allocations in the commercial fishery. The recreational fishery regulations include overall recreational harvest limits, minimum size, seasons, and bag limits.

COD SWIMBLADDER FUNCTIONS

In cod the swimbladder is used as a source of sound production. Rapid contraction of extrinsic swimbladder muscles against the swimbladder wall causes the sound-producing vibrations. These low frequency sounds occur during complex mating behavior as well as during part of aggressive displays. Those threat displays may also be accompanied by raised fins, flared gill covers, and an open jaw.

The swimbladder also serves to detect sound from a variety of sources, for example, other cods or predators such as toothed whales. Those sound waves cause the bladder to vibrate. Anterior extensions of the bladder extend close to the fish's inner ear where ear stones (otoliths) within the inner ear's chambers are displaced and affect the sensory hair cells lining the chambers that ultimately transmit information to the auditory nerve.

Cod larvae live in the plankton for the first 2 months after which the cod is primarily a bottom-living species although it ranges from the surface to depths that may exceed 1000 ft. To ascend from great depth, the cod's swimbladder must reduce its gas pressure. If not, the bladder would increase in volume and burst. To descend, the pressure within the bladder must be increased. A gas gland built into the wall of the bladder assists in pumping gases into the bladder.

LITERATURE. Arnold and Walker 1992; Pelster and Scheid 1993; Rowe and Hutchings 2004

Atlantic tomcod

Microgadus tomcod

Other common names: frostfish
Labrador to Virginia
Max. length = 17 in., av. = 10 in.

FIELD CHARACTERISTICS. Unlike the Atlantic cod, the tomcod has a rounded tail, a small chin barbel, and a filamentous pelvic fin. The tomcod's olivaceous mottled body pattern is well adapted to the shallow benthic environment of our local bays.

ECOLOGY/LIFE HISTORY. The tomcod is a year-round resident of our bays. During the winter, it survives and can spawn in icy conditions owing, in part, to the antifreeze proteins that tomcods produce seasonally. This species might also be found in rivers. In the Hudson River, the tomcod is anadromous and moves upstream to brackish waters to spawn during November through February.

Tomcods do not have a long life span. Some populations may be reproductively mature at the end of their first year and live for only an additional year. Four years appears to be the maximum age for this species.

These fish are opportunistic feeders. Although they may prefer small crustaceans, they also eat marine worms and a variety of small fishes. Because of their relatively small size and availability, tomcods are preyed upon by a wide range of fish-eating predators.

The origin of their common name is unknown although locally they may be called tommy cod or frostfish. The latter name is unhelpful because silver hake and whiting are also frequently called frostfish. The genus, *Microgadus*, is appropriate since it means "small cod." *Microgadus promixus*, the Pacific tomcod, is the only other species within this genus.

LITERATURE. Reisman et al. 1984

Haddock

Melanogrammus aeglefinus

Newfoundland to New Jersey and in the eastern Atlantic
Max. length = 3.5 ft., common = 1.5 ft.

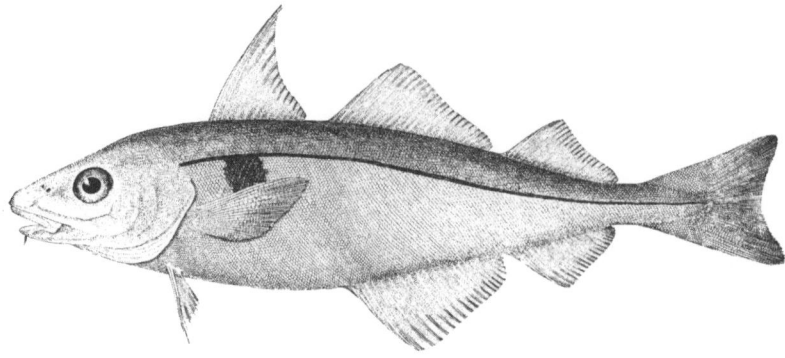

FIELD CHARACTERISTICS. The haddock has a single dark black blotch just above the pectoral fin and the lateral line is accentuated by a dark stripe. The genus name, *Melanogrammus*, means "black line" in Greek. The first dorsal fin is tall and pointed rather than rounded as is the case in the cod, tomcod, and pollock.

FISHERIES. Although an important commercial and recreational fishery in New England, haddock has been less important in New York, particularly during the past 30 years due to abundance issues, shifting distributions patterns, as well as changes in management that cut many New York commercial fishermen out of the haddock fishery. New York commercial landings of haddock were highest during the late 1930s and early 1940s when they ranged between 7 and 11 million pounds. Since the mid-1980s, commercial haddock landings in New York have been in the range of tens of thousands of pounds with no reported landings in some years. Other than a few fish in 2019, there have been no reported recreational landings of haddock in New York for the past 40 years.

MANAGEMENT. Haddock is managed jointly by the NEFMC, NMFS, and the NYSDEC. The NEFMC and NMFS manage haddock as part of the Northeast Multispecies (groundfish) FMP (NEFMC 1985), which also includes a variety of other species including cod, windowpane, and winter and yellowtail flounder. The fishery is managed as two separate stocks: Georges Bank and Gulf of Maine. Haddock from the Georges Bank stock are occasionally found off New York and Rhode Island, and that stock is not overfished nor is it subject to overfishing (U.S. Department of Commerce 2021). The commercial fishery is managed by overall quotas, minimum size, seasons, gear restrictions, trip limits, and individual allocations. The recreational fishery regulations include overall recreational harvest limits, minimum size, seasons, and bag limits. New York has a minimum commercial and recreational minimum size.

Pollock

Pollachius virens

Other common names: American pollock, coalfish, Boston bluefish, saithe
Greenland to North Carolina and the eastern Atlantic Ocean
Max. length = 4.2 ft.

FIELD CHARACTERISTICS. The pollock has a forked tail, a lower projecting jaw, and an olive-green color above a very straight pale lateral line. The species name *virens* means green in Latin. Below the lateral line, the color ranges from gray to silvery, and the ventral surface is light. Countershading, dark dorsal and light ventral, is common in fishes that swim in midwater or near the surface. Although the cod family members usually have chin barbels, the pollock's barbel is barely discernable or occasionally is absent.

ECOLOGY/LIFE HISTORY. Pollock prefers cold water, and dense schools migrate along the coast to and from midwinter spawning grounds. The role of the lateral line in maintaining school integrity was demonstrated by experiments with pollock whose vision was temporarily obscured. When these fish were introduced into a school of intact pollock, schooling continued unimpaired. However, when the lateral line was cut near the gill cover, these "blinded" pollock failed to school successfully.

This fish shares many life history qualities with the Atlantic cod. For example, the pollock is very fecund, uses its swimbladder for sound production, and also has a similar distribution, habitat, diet, and range of predators.

In general, warmer tropical waters have more species diversity but temperate and cold-water fishes contain the greatest density of fishes and, as such, contain the most important food fishes (e.g., cod, pollock, haddock, and a variety of flatfish and herring family members). The largest food fishery in the world is the North Pacific walleye pollock.

LITERATURE. Pitcher et al. 1976; Montgomery et al. 1997

CUSK-EELS
Family Ophidiidae (*brotulas y congriperles*)

This is a large family (222 species) and occurs within all the major oceans and at varying depths. In total, 16 species occur along the Atlantic coast of the United States. Their bodies are elongate and eel-like, except unlike eels, this family has pelvic fins. The small pelvic fins in cusk-eels are somewhat unusual in that they are placed on the throat region and look very much like barbels although this family has no barbels. Like eels the long dorsal and anal fins are continuous with the caudal fin.

Striped cusk-eel

Ophidion marginatum

Long Island to Gulf of Mexico
Max. length = 9 in.

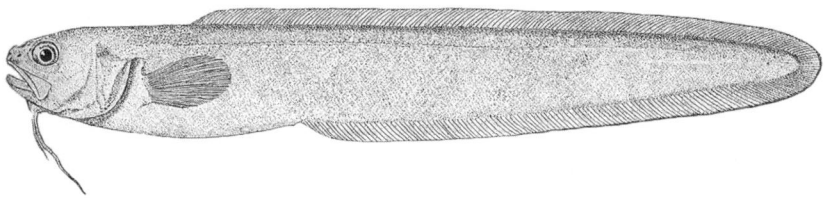

FIELD CHARACTERISTICS. It is unlikely that the cusk-eel would be confused with the cusk, a member of the cod family that has only a superficial similarity to the cusk-eel. The most obvious feature of the striped cusk-eel is the two to three dark lateral stripes along the entire length of the body. This species tends to be more active at night but otherwise lives in a burrow made in the sediment. A special sound-producing structure associated with muscles attached to the cranium is used during reproductive activity.

LITERATURE. Schwartz 1997

TOADFISHES
Family Batrachoididae (*peces sapo*)

This family contains 73 species broadly distributed along the world's coasts. Nine species occur along the Atlantic coast but only one is encountered locally. All family members are benthic, possess a large pair of pectoral fins, and their smaller pelvic fins are placed in a jugular position, in front of the pectorals. Family members have two to three dorsal spines preceding the long, soft-rayed, dorsal fin. There may also be one or two opercular spines on the gill cover (opercle).

Oyster toadfish

Opsanus tau (sapo)

Maine to Florida
Max. length = 15 in.

FIELD CHARACTERISTICS. The body is robust and scaleless, the head is broad and depressed. Fleshy cirri (flaps) are arranged above the eyes, along the margins of the lower jaw, and elsewhere. These features in conjunction with the cryptic olivaceous to black body markings obscure the presence of the fish especially on sandy/muddy backgrounds or if in algae or eelgrass beds. While collecting scallops in eelgrass, it is not uncommon to mistake the brown head of a toadfish for a large scallop, which leads to an alarming moment for both fish and far-sighted scalloper.

Opsanus (looking upward in Greek) and *tau* (the letter T) refers the position of the eyes and the T-shape formed by the bones of the head when dried.

TOADFISH SOUND PRODUCTION

Many fishes produce sound. Toadfishes, among other species, possess a pair of sonic muscles attached to the sides of the swimbladder. When those muscles contract the bladder vibrates. The swimbladder and the sonic muscles are larger in males than females. Breeding male toadfish produce long-duration "boat-whistles" lasting 500–700 milliseconds. These sounds are produced intermittently for many hours during the mating season and function to attract a female and possibly permit the female to select the proper and best mate. Boat-whistles are made most frequently by nest-guarding males without eggs rather than with eggs. The sonic muscles in these fish have one of the fastest contraction cycles in the animal kingdom. Adaptations at the structural and biochemical level promote increased speed of contraction and fatigue resistance. Short duration grunts of up to 200 milliseconds may be produced by both sexes throughout the year. It is suggested that grunts function during aggressive encounters. The fish can be aggressive when confronted and will ambush prey. The fish is most conspicuous during its summer breeding season when females lay a clutch of large (5 mm) adhesive eggs in the nests established by males. Males guard the eggs for about a month. Those nests are usually natural depressions in the bottom although discarded tires or empty plastic containers may also be occupied as nests.

LITERATURE. Rome et al. 1996; Rome and Lindstedt 1998

ECOLOGY/LIFE HISTORY. The accepted common name probably was inspired by their toad-like appearance, ability to make sound and their common occurrence in oyster beds, where they feed on those oysters, other mollusks, as as on crustaceans and small fishes.

Toadfish has not generally been considered a food fish, but it is gaining some popularity with consumers. Further, oyster toadfish has been successfully tested to be used for the biological control of crab predation on small hard clams. The fish's large mouth with strong, blunt teeth makes it an ideal predator upon crustaceans such as mud and blue crabs. When toadfish were placed among small seed clams, the fish reduced the number of crabs or at least reduced crab feeding efficiencies, and as a result clam survival increased. However, experimental field studies in Chesapeake Bay suggest that the toadfish is not effective in controlling oyster spat predators.

LITERATURE. Bisker and Castagna 1989; Abbe and Breitburg 1992

GOOSEFISHES
Family Lophiidae (*rapes*)

This family contains four genera and 25 species that are generally found in cold water, including the Arctic and portions of the Atlantic, Pacific, and Indian Oceans. It belongs to a large order of marine fishes (Lophiiformes) with over 300 species. These fishes include deep-sea anglerfishes, frogfishes, and batfishes that are all characterized by possessing a mobile angling apparatus that includes a rod-and-line-like "illicium" and a lure "esca" used to attract prey. This unusual fishing device develops from the first ray of the spiny dorsal fin.

Goosefish

Lophius americanus

Other common names: monkfish, American angler, allmouth, molligut
Newfoundland to northern Florida
Adult length from 2 to 4 ft.

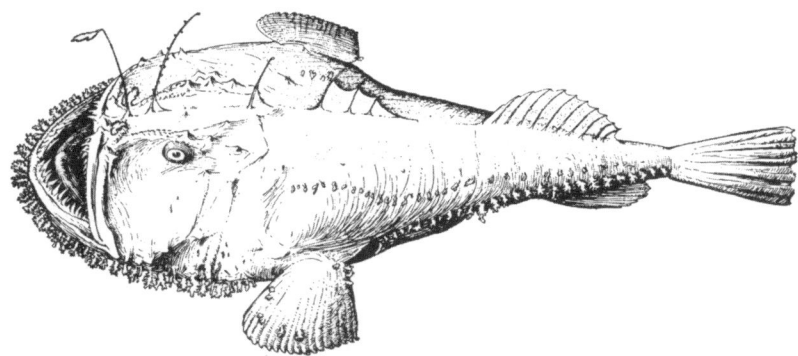

FIELD CHARACTERISTICS. As a lie-in-wait benthic predator, the goosefish is darkly colored and has a fringe of small flaps around a large projecting jaw. Somewhat uniquely, there are no typical gill covers that are in front of the pectoral fins. Instead, there is a small gill opening behind each pectoral fin base.

ECOLOGY/LIFE HISTORY. The goosefish ranges from shallow water in bays to over 2000 ft. depths offshore. As in many broad ranging but generally northern fishes, the goosefish is uncommon south of North Carolina.

This is a fish with many unusual anatomical and life history features. For example, females deposit over a million eggs in a thin, free-floating gelatinous mass secreted by the ovaries. These rafts can be up to 30 ft. long and nearly 3 ft. wide.

The fish is inactive until the instant in which, via sudden and rapid suction feeding, it engulfs the prey. The prey, commonly, are smaller fishes attracted to the lure, but the goosefish has also been known to feed on diving seabirds. Large prey is accommodated by the wide mouth, containing well-developed, recurved, depressible teeth. The pectoral fins are arm-like and may facilitate the ability to maintain the motionless position needed as the fish lies concealed on the bottom. In general, a fish's unique behavior is an adaptive product of its anatomy and physiology. The goosefish and other relatively inactive fishes, like toadfishes, have reduced metabolic demands. As expected, these fishes have been shown to have a lower blood oxygen capacity and a smaller gill area per body weight when compared with more active fishes such as mackerel and menhaden.

LITERATURE. Gosline 1996

FISHERIES. In New York and elsewhere, goosefish are frequently marketed under the name of monkfish. Goosefish was long considered an underutilized species as market demand, and thus landings, were low through most of the twentieth century. During that time landings in New York averaged less than 50,000 lb./yr. and were an incidental catch when fishing for other species. In the late 1970s landings started to increase substantially in response to the expanding domestic market and export opportunities primarily in Europe and Asia. New York landings peaked at 2.5 million pounds in the late 1990s and have averaged 1.5 to 2.0 million pounds since then. Goosefish are also a very valuable fish with a high price paid to the fishermen. Goosefish have become one of the most valuable finfish landed in New York and in the Northeast. The head is very large and constitutes a significant portion of the body. However, goosefish have a very large meaty tail, and it is the tail portion of the fish that is eaten. The tail meat is firm, dense, relatively boneless, and mild-tasting. In Europe and Asia, the livers are also in high demand and a U.S. export market has developed to help satisfy that demand. Goosefish are caught primarily with specially designed gillnets and to a lesser extent with trawls. Fish are sometimes landed whole and sometimes just the tails and livers. There is essentially no recreational fishery for goosefish as they will not take bait or artificial lures: goosefish use their large mouth to consume and eat only live whole fish.

MANAGEMENT. Goosefish is managed sustainably by the NMFS, NYSDEC, NEFMC, and the MAFMC through the Monkfish (Goosefish) FMP (NEFMC 1998). The NEFMC has the lead on the Goosefish FMP. The commercial fishery is divided into northern and southern management areas with southern George Bank dividing the two. Federal management measures include an overall quota for each area; limited number of fishing days assigned to each permit; trip limits; and minimum fish size. The goosefish stock has been rebuilt and overfishing is not occurring (NMFS 2020). There are no management measures in the FMP for recreational fishermen.

New York has a commercial and recreational minimum size regulation in effect for both whole fish and tails. As mentioned above, goosefish are sometimes landed whole and sometimes just the tails and livers. There is also a limit on the percent of the commercial goosefish catch that can be livers in order to prevent landing just livers and discarding the rest of the fish.

MULLETS
Family Mugilidae (*lisas*)

The family is distributed worldwide in temperate and tropical coastal seas as well as in brackish and freshwater habitats. There are 17 genera and 72 species within the family, and in many parts of the world, the mullet is considered a food fish and an object of aquaculture. In North America there are 12 species, but only 2 are found in the Northeast and Mid-Atlantic region. In general, the fish is cylindrical in shape and has a short snout, small mouth with small teeth, if any, and only a faint lateral line, if there is one at all. More easily noted are the widely separated two dorsal fins.

Mullets are herbivores and swim in large schools feeding on benthic algae, diatoms, and detritus. Their gill rakers are designed to filter out large food particles that are then directed to the muscular, gizzard-like stomach. This latter structure further breaks down the plant material and detritus. In herbivorous fishes, digestion is enhanced by increased length of the intestines. In mullets that length could be nearly eight times the fish's body length.

Striped mullet

Mugil cephalus (lisa rayada)

Other common names: black mullet, gray mullet
Nova Scotia to most parts of the Caribbean to Brazil, and worldwide
Max. length = 25 in.

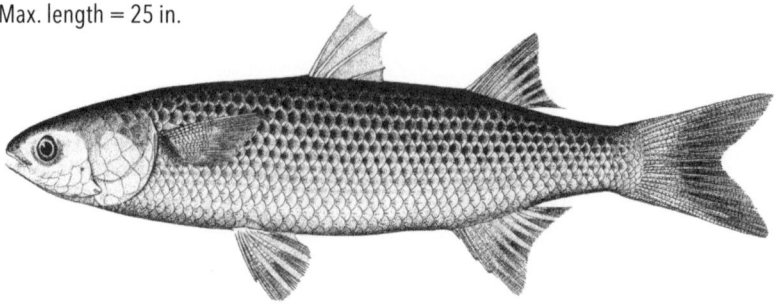

FIELD CHARACTERISTICS. At a size larger than about 3 in., the striped mullet has scales on its side marked with a dusky blotch to form horizontal stripes. Other less-conspicuous distinctions between the striped and the white mullet below involve the number of soft rays in the anal fin (11 vs. 12), the number of lateral line scales (37–43 vs. 33–39), the size of a dark spot at the base of the pectoral fin, and certain gold pigmentation features of the operculum and iris.

Both local species of mullets have adipose eyelids. This is not unique to this family and can be found in other fishes, for example, herrings, jacks, tunas, and salmon. These coverings are made of transparent skin. The eyelid has a small aperture or slit and serves to protect the eye as well as contributes to streamlining the head.

ECOLOGY/LIFE HISTORY. Although locally found in coastal waters, it is common in freshwater elsewhere; for example, in Texas and Louisiana it can be found in rivers but, in general, is not a permanent resident of freshwaters.

White mullet

Mugil curema (lisa blanca)

Other common names: silver mullet
Massachusetts through the Caribbean to Brazil and also in the eastern Pacific Ocean
Max. length = 14 in.

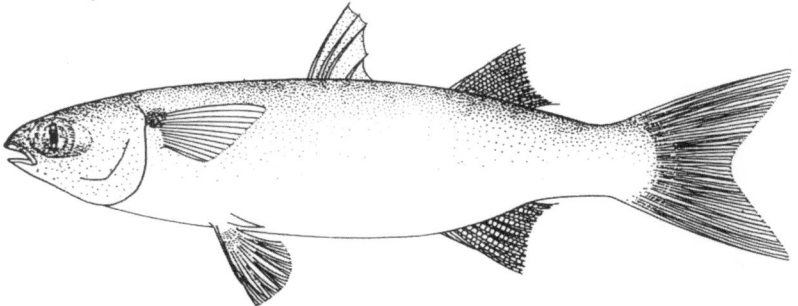

FIELD CHARACTERISTICS. *Curema* originates from the Portuguese name for Queriman = Surinam, a country on the northeastern Atlantic coast of South America. In contrast to the striped mullet, the white mullet tends to prefer more saline conditions and thus would not often be found in freshwater habitats.

NEW WORLD SILVERSIDES
Family Atherinopsidae (*charales y pejerryes*)

This is a widely distributed family with 11 genera and 108 species. Representatives can be found throughout the New World, that is, North, Central, and South America, in marine or freshwater habitat. Formerly, the New World silversides were not distinguished from those found in the Old World, and both were classified as members of a single family, the Atherinidae.

Rough silverside

Membras martinica (pejerrey rasposo)

New York to Yucatan, Mexico
Max. length = 5 in.

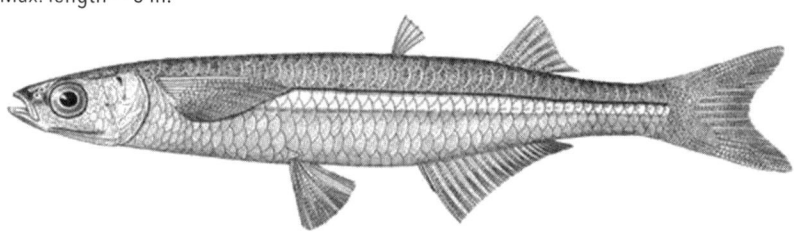

FIELD CHARACTERISTICS. The common name recognizes the most distinctive feature of this silverside, namely, that the body scales are rough to the touch. A further distinction is that the bases of both the dorsal and anal fins are covered with large scales. In general, the habitat of the rough silverside tends to be less associated with salt marshes than the other New York silversides. The most common habitat is coastal water, but throughout its range it can also be found in low-salinity environments.

Although the common name is informative, the species name (*martinica*) is less so. The name was given to this species by the distinguished French naturalist, Achille Valenciennes in 1835. He did so based on the curious assignment of Martinique (in the Lesser Antilles of the Caribbean) as the typical location for this species, for which there is no evidence.

TELEOSTEAN EVOLUTION

Teleosts, a division of bony fishes, have evolved over more than 200 million years, and as they evolved from ancestral to more advanced families, certain trends in the condition of some anatomical characters generally occurred. Silversides appear to have a set of features that represent an intermediate condition between that which characterizes ancestral teleostean families and the more advanced condition. Ancestral features possessed by silversides include smooth-edged cycloid scales rather than ctenoid scales that have comb-like teeth on their edges and pelvic fins that are more abdominal than thoracic. On the other hand, more evolutionarily advanced features of this family seem to predominate. Some examples are the two dorsal fins, the first possessing spiny rays; pectoral fins placed on the midline rather than ventrally located; and a pelvic fin formula of I,5 (a single spiny ray followed by five soft rays). Further, the protractile premaxilla bone excludes the maxilla bone from participating in the mouth's gape and the swimbladder is not connected to the gut (physoclistic condition).

Inland silverside

Menidia beryllina (plateadito salado)

Gulf of Maine to Mexico
Max. length = 3.8 in.

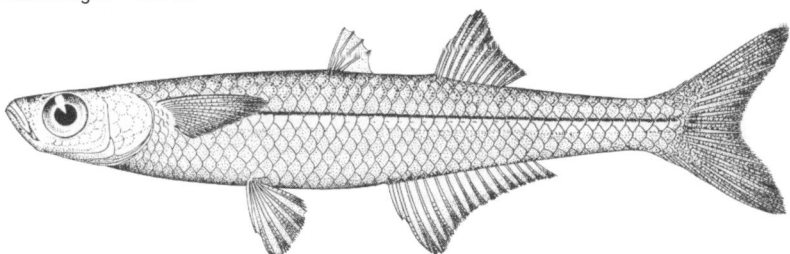

FIELD CHARACTERISTICS. The inland silverside is found in less saline and more brackish parts of the estuary than the Atlantic silverside. In *M. menidia* the first dorsal fin lies at the end of an imaginary vertical line from the position of the anus to the dorsal area. In contrast, in *M. beryllina* the first dorsal fin originates well in front of that vertical line from the anus.

Atlantic silverside

Menidia menidia

Other common names: whitebait, spearing, rain minnow, green smelt, sand smelt
Nova Scotia to northern Florida
Max. length = 4.3 in.

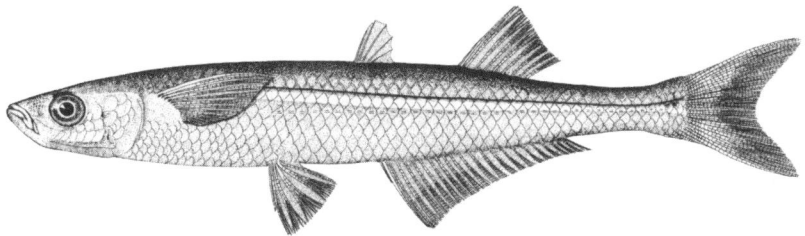

FIELD CHARACTERISTICS. Usually found in shallow water adjacent to marine wetlands or beaches. As in all members of this family, they have two widely separated dorsal fins and a conspicuously silver lateral stripe.

ECOLOGY/LIFE HISTORY. Atlantic silversides are gregarious fishes that aggregate into large schools. As such these fish are important prey organisms for bluefish, striped bass, terns, cormorants, and other predators. They are commercially collected as bait fish. Young of the year bluefish exploit the abundance of small silversides in the spring. Bluefish spawn offshore in southern waters in the spring and are recruited into New York's bays just at the same time that a rich supply of food in the form of small silversides is available. For that reason, young bluefish have one of the fastest growing first years of any fish species. The silversides feed on a variety of small invertebrates and zooplankton that are associated with marine wetlands. Some Atlantic silversides migrate during the winter to offshore waters where they are prey for large piscivores that do not occur in shallow, marsh-bordered areas. In this way, the energy acquired by silversides from salt marsh communities is transferred to open water.

Silversides follow a lunar rhythm during the spring and summer spawning season. These fish spawn on a new and full moon when the tides are at their highest. Eggs are deposited at the bases or fronds of smooth cordgrass, *Spartina alterniflora*, the dominant grass found on the margins of the tidal creeks and bays. In two weeks, during the next high tide, the eggs, which are resistant to drying, will be agitated, and the young fry will hatch from the egg. A comparable spawning behavior occurs in the related California grunion, *Leuresthes tenuis*. In that case, at high tide, females swim as far up the sandy beach they can, then using their tail, excavate a sandy burrow where the eggs are laid and simultaneously fertilized. The eggs hatch about 10 days later.

During larval development in fishes, the sex of the individual is determined by major sex-determining genes. However, in silversides, the sex is determined by the joint effects of genes and temperature. It was concluded that young produced in the bays and estuaries during the spring experience low temperatures that cause most larvae to become females. High temperatures that occur during the summer cause most silverside larvae to differentiate into males. The advantage is clear. Females benefit more than males do from a longer growing season. Females developing early in the season can feed for a longer period and thus be larger, in general, than males. A larger female is more fecund, but larger males are not significantly more fertile than smaller ones.

LITERATURE. Walker 1959; Conover and Kynard 1981

HALFBEAKS
Family Hemiramphidae (*pajaritos*)

There are 12 genera and 109 species within this family. They are widely distributed and are generally found in warm temperate to tropical marine waters although there are some freshwater species.

Halfbeaks are distinguished by having a lower jaw longer than the upper one. They are surface-dwelling, schooling fishes that are generally herbivorous and feed on sea grasses and some plankton. The elongate jaw may function to direct food into the mouth.

In some halfbeaks the lower lobe of the caudal fin is slightly enlarged and so are the pectoral fins, suggesting a relationship with the flyingfish family, although in most halfbeaks the pectoral fins are not nearly as large as in flyingfishes. Further, halfbeaks frequently jump out of the water and skip over the surface. For these reasons, some taxonomists classify halfbeaks and flyingfishes in the same family.

False silverstripe halfbeak

Hyporhamphus meeki (agujeta flaca)

Other common names: Meek's halfbeak, American halfbeak
Inshore New Brunswick, Canada to the Yucatan, Mexico
Max. length = 7 in.

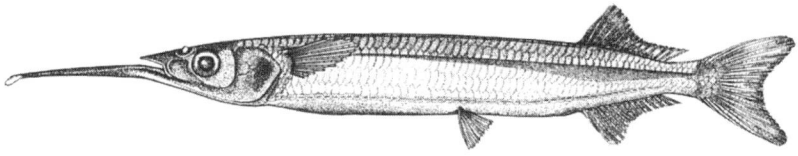

FIELD CHARACTERISTICS. Consistent with its unique lower jaw, the genus *Hyporhamphus* is Greek for "lower beak" and the species *meeki* is named after Seth E. Meek, a nineteenth–early twentieth century Field Museum of Natural History ichthyologist.

The body is elongate and compressed. It has a silvery band running along its side, and the long lower jaw has a red fleshy tip. The dorsal and anal fins are located posteriorly, and the caudal fin is moderately forked.

ECOLOGY/LIFE HISTORY. Like other generally warm-water species, this wide-ranging inshore fish occurs more commonly in the southern part of that range. Halfbeaks are preyed upon by many large gamefish, and as one might expect, this species is used as baitfish by anglers.

NEEDLEFISHES
Family Belonidae (*agujones*)

The family includes 10 genera and 34 species. Most are found along temperate and tropical marine coasts throughout the world, although 12 of the species are found in freshwater.

The most distinguishing characteristics of this family are the upper and lower jaws that extend into long beaks each with numerous, fine, sharp teeth. As juveniles, needlefishes have relatively short jaws of equal length. As the fish grows, the lower jaw elongates first, resulting in a "halfbeak" stage, and then the upper jaw grows to equal the lower one in length. While in the early stage, the diet is primarily large zooplankton but after full development of the beaks, these fish are exclusively fish eaters.

Atlantic needlefish

Strongylura marina (agujon verde)

Other common names: silver gar, saltwater gar
Coasts of Maine to Brazil
Max. length = 2 ft.

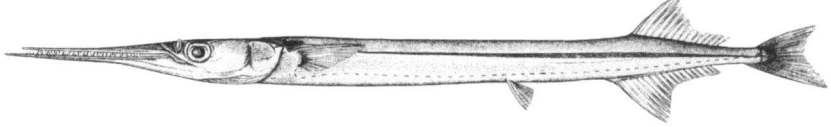

FIELD CHARACTERISTICS. Considering the combination of very long upper and low jaws with many sharp teeth, the "needlefish" is perfectly named. The compact, elongate body and posteriorly placed dorsal and anal fins contribute to the thrust used to catch swimming prey.

ECOLOGY/LIFE HISTORY. This species is a surface-dwelling lie-in-wait predator that uses its narrow profile, silvery sides, and counter-shaded body to remain undetected until prey are within striking distance.

SAURIES
Family Scomberesocidae (*papardas*)

There are only four species within this family, and two are found in the Atlantic. These are long and slim fishes, and some species have elongated jaws. Because of their jaws, these fishes have occasionally been included within the needlefish family. The single dorsal and anal fins are placed far back on the body, and a series of small finlets are arranged between those fins and the caudal fin.

Atlantic saury

Scomberesox saurus

Newfoundland to Florida, eastern Atlantic and Mediterranean
Max. length = 2.5 ft.

FIELD CHARACTERISTICS. This is the largest species of saury. It swims in schools at the surface and may be confused with halfbeaks or needlefishes. However, halfbeaks have only one (the lower) jaw elongated and needlefish, which have both long jaws, do not have a set of five or six finlets in back of the dorsal and anal fins.

KILLIFISHES
Family Fundulidae (*sardinillas*)

Of the 41 North and Central American species within this family, most live in eastern North America. Many occupy inland freshwater habitats, and 12 occur along the Atlantic coast. Five of those can be found in New York's waters. Members of this family are an important part of the diet of larger commercially important fishes, such as summer flounder, bluefish, and striped bass. Egrets and herons also prey upon these killifishes.

Most killifishes live in predictable local habitats. The mummichog and the striped killifish can be found in salt marsh bordered tidal creeks. The rainwater killifish would be in ponds that are more brackish. The spotfin killifish is the least abundant and smallest *Fundulus* and when found would likely occur in shallow, brackish water. The banded killifish, *F. diaphanus*, is found in only freshwater so is not reviewed here.

Mummichog

Fundulus heteroclitus

Other common names: killi
Newfoundland to northern Florida
Max. length = 5.5 in.

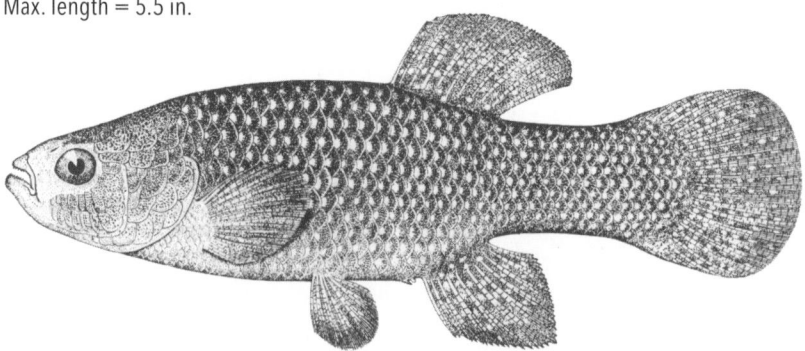

FIELD CHARACTERISTICS. The length of the snout is equal to the diameter of the eye. The mouth is terminal, the caudal peduncle is deep, and the caudal fin is round. Males may be more colorful during the breeding season, but other than having about 15 thin dusky to pale bars, both sexes can be somewhat nondescript.

ECOLOGY/LIFE HISTORY. These are ubiquitous, small, schooling fish found in tidal waters adjacent to salt marshes. The name mummichog originates from the Narragansett *moamitteaug*, which means "going in crowds." Indeed, they are the most abundant fish adjacent to salt marshes in our region. In a Delaware Bay marsh, the mummichog density was equal to 20 per square meter.

Mummichogs spawn in the spring and summer and deposit eggs in empty mussel shells, on algal mats at the base of the dominant salt marsh cordgrass (*Spartina alterniflora*), or on the blades of that plant. This activity occurs during full or new moon when the tide is high. This timing strands the eggs, which tolerate dry conditions, until the next high tide in approximately two weeks. At that time, upon inundation, the eggs are stimulated to hatch, and the larvae develop within the protection of the salt marsh.

The mummichog is an extraordinarily tolerant fish. It can survive a wide range of water temperature, salinity, and dissolved oxygen. This tolerance is valuable for an estuarine fish that lives throughout the year in shallow, tidally influenced habitats from Canada to Florida. This fish and related species have been reported to occur naturally over a wide range of salinities. In the laboratory, these fishes have shown considerable salinity tolerance with upper limits that are over three times the 35 ppt salinity of seawater. This capacity to be tolerant in general is also adaptive at those times when they are exposed to toxic chemicals that enter nearshore waters adjacent to industries, agriculture, marinas, and golf courses. It has been shown that mummichog embryos from populations that live in contaminated waters develop some resistance to toxicants, for example, methyl mercury.

Spotfin killifish

Fundulus luciae

Coastal waters from Massachusetts to Georgia
Max. length = 2.0 in.

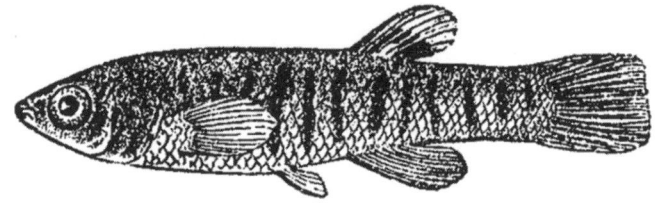

FIELD CHARACTERISTICS. This is the smallest killifish. The common name applies to the male, which possesses a distinct black spot on the posterior edge of the dorsal fin although this is only prominent during the breeding season. A distinction of the spotfin killifish is the small number of fin rays (8–9) in its dorsal fin, which makes the base of the dorsal fin shorter than the base of the anal fin. As in all killifish, males are smaller than females.

Striped killifish

Fundulus majalis

New Hampshire to the Gulf of Mexico
Max. length = 7 in.

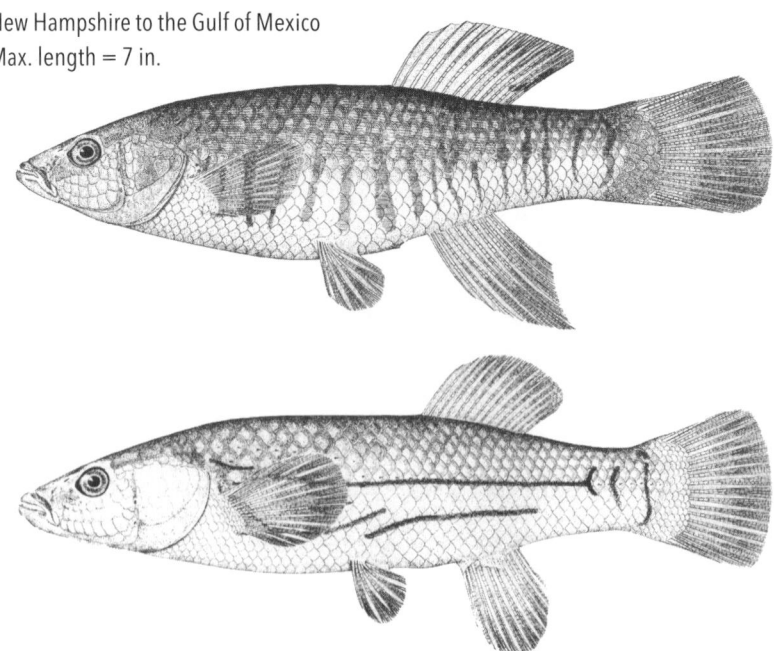

FIELD CHARACTERISTICS. The two species that would most commonly share a habitat are the striped killifish and the mummichog. The striped killifish is distinguished by the bold markings (black stripes in females and many dark bars in males). Both sexes have a relatively long snout. The distance between the tip of the snout and the front of the eye is longer than the diameter of the eye. In contrast, that distance is equal to the eye's diameter in the mummichog.

ECOLOGY/LIFE HISTORY. As in most killifishes, these are schooling fish. The habitat of the striped killifish overlaps with the mummichog, although the former tends to be found in higher salinities and thus in slightly more diverse habitats. In local waters, breeding tends to begin in May. Appropriately, the species name *majalis* is Latin for May. Although most members of the killifish family have common names that identify them as killifish, the close relationship between the mummichog and the striped killifish is obvious only when one knows that the two fishes share the same genus.

Rainwater killifish

Lucania parva (sardinilla de lluvia)

Cape Cod to Florida
Max. length = 2.7 in.

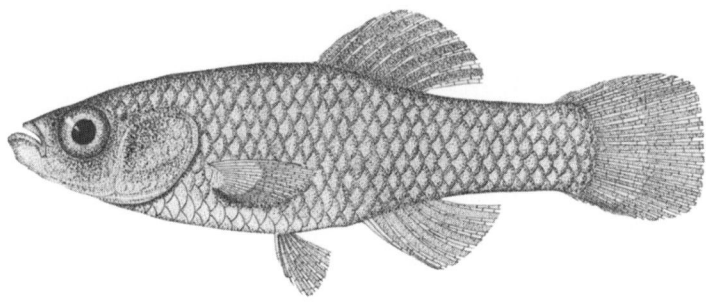

FIELD CHARACTERISTICS. These are small fish. The species name *parva* is Latin for small. Males have a black spot on the anterior of the single dorsal fin. The rear tip of the pectoral fin is barely beneath the origin of the dorsal fin.

LITERATURE. Weis and Weis 1977; Weis 2002

PUPFISHES
Family Cyprinodontidae (*cachorritos*)

The pupfish family has nine genera and 104 species ranging from United States, Mexico, Central America, the Caribbean, and northern South America to parts of the Mediterranean region. The genus *Cyprinodon* has about 40 species. Typical pupfish habitats are found in the freshwaters of Mexico, where 34 different species of *Cyprinodon* occur. Nine other freshwater *Cyprinodon* species occur in the United States, but only one species of *Cyprinodon* is found in New York.

Sheepshead minnow

Cyprinodon variegatus (bolin)

Massachusetts to Mexico
Max. length = 3 in.

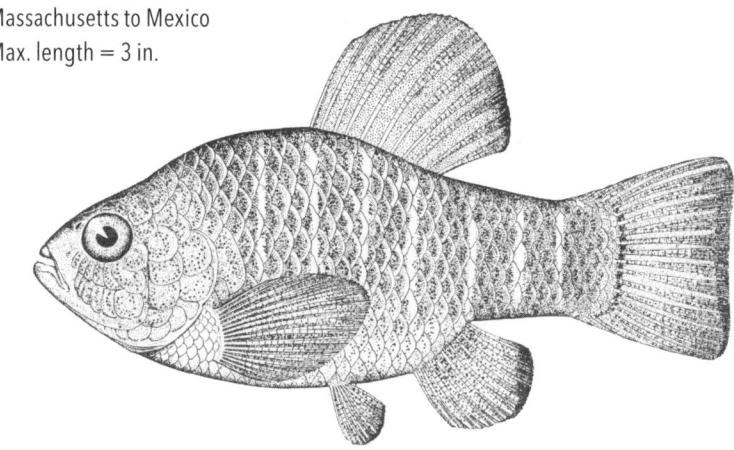

FIELD CHARACTERISTICS. This species occurs in brackish and coastal waters and is most frequently found living sympatrically with mummichog killifish. In both mummichogs and sheepshead minnows, the origin of the dorsal fin is anterior to the origin of the anal fin, but in the sheepshead minnow that is more apparent. The sheepshead minnow can also be distinguished from the mummichog by the sharp tricuspid teeth in the jaws of *Cyprinodon* rather than the conical teeth of killifishes. Most conspicuously, the sheepshead minnow has a distinctively robust, deep body so that the maximum depth of the body is equal to half the fish's standard length.

Incomplete bars and variegated markings on the body of individual specimens gave rise to the species name *variegatus*. Within the breeding season, males are colorful with a bluish dorsal and an orange abdomen and with pectoral, pelvic, and anal fins also being orange.

ECOLOGY/LIFE HISTORY. This is a common member of the salt marsh fish community. The sheepshead minnow is an important member of the food chain. This fish primarily feeds on detritus and small invertebrates, and is prey for larger piscivorous fishes and birds.

The common name "sheepshead minnow" is misleading. The true minnow and carp family, Cyprinidae, is a very different, unrelated, primarily freshwater family and does not tolerate high salinities. In contrast, the sheepshead minnow has remarkable osmoregulatory abilities. Populations of this species have been found in hypersaline lagoons containing salinities of over four times that of full seawater. A further distinction is that members of the minnow family, such as carp, are toothless. In fact, the name *Cyprinodon* is Greek for "carp tooth/ toothed carp" and refers to the fact this species has teeth unlike a carp.

LITERATURE. Martin 1972; Nordie 1985; Hagan et al. 2007

DORIES
Family Zeidae (*peces de San Pedro*)

There are five species in the family. They occur in the benthopelagic and meso-pelagic (twilight zone) from 660 to 3300 ft. in the Atlantic, Pacific and Indian Oceans.

Buckler dory

Zenopsis conchifera

Other common names: American John Dory,
    St. Peter fish
Nova Scotia to Gulf of Mexico to Brazil,
    eastern Atlantic and Indian Ocean
Max. length = 2.4 ft.

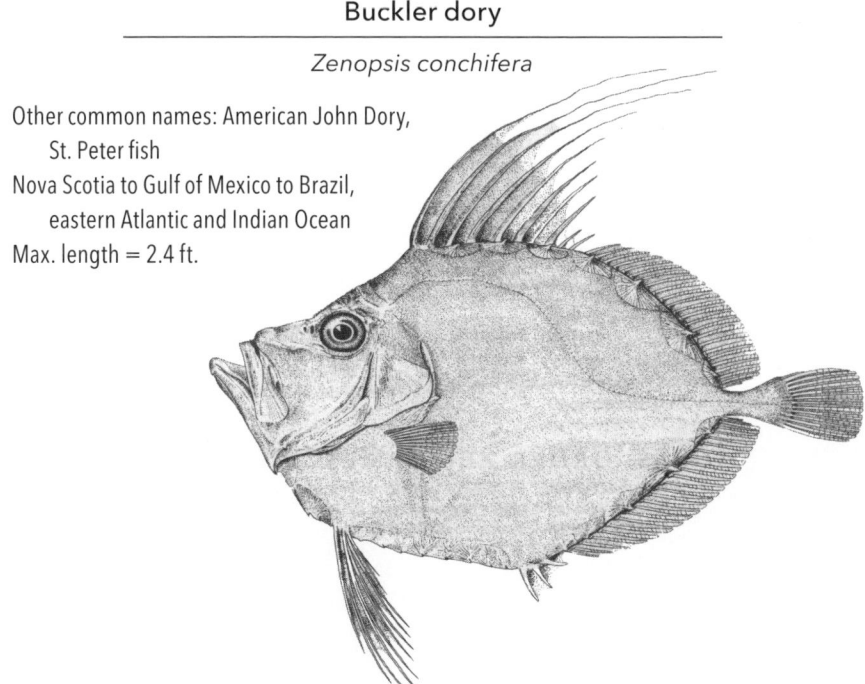

FIELD CHARACTERISTICS. This dory has large scales (bucklers) with posteriorly directed processes at the base of the dorsal and anal fins. Scutes are present along the belly.

The body is very compressed, deep, and silvery with a grayish blotch in the center well below a highly arched lateral line. The mouth is large and protrusile, and there are long anterior dorsal spiny fin rays with filamentous ends.

STICKLEBACKS
Family Gasterosteidae (*epinoches*)

Of the five genera and minimally seven species in this family, three genera and four species can be found on Long Island and the lower Hudson. Depending upon the species, members of this family may be found in marine, brackish, or freshwater habitats. The number of dorsal spines varies with the species from three in *Gasterosteus,* four in *Apeltes*, and nine in *Pungitius*. Sticklebacks lack scales although in some populations of *Gasterosteus aculeatus*, bony lateral plates protect the flanks. Males of all members of this family build nests within which eggs are placed, fertilized, and guarded until the fry hatch. During the period of parental care, some males aerate the eggs by fanning with their pectoral fins. Both species of *Gasterosteus* build the nests on the substratum whereas *Apeltes* and *Pungitius* build nests in vegetation above the bottom. Within the stickle-back family, each species has a distinctly different male courtship dance and nuptial coloration. In Maine, where all four species can share the same shallow nearshore coastal habitat, differences in behavior and male coloration would likely serve to maintain reproductive isolation within the family.

Fourspine stickleback

Apeltes quadracus

Other common names: bloody stickleback
Newfoundland to Virginia
Max. length = 2.5 in.

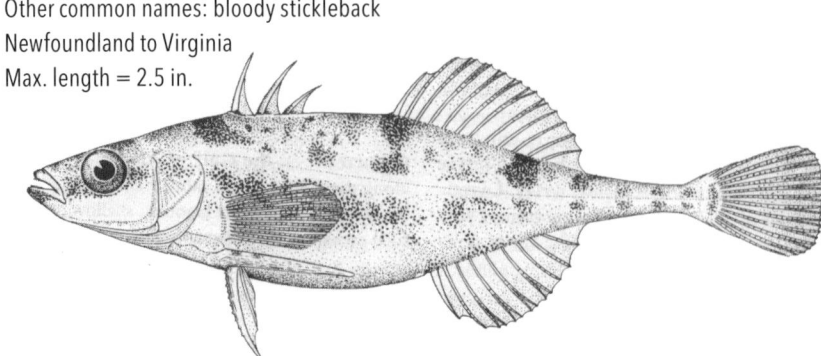

FIELD CHARACTERISTICS. This species is found in coastal and estuarine waters and is, with the threespine stickleback, one of the most common sticklebacks on Long Island. *Apeltes* most distinctive feature is the four free dorsal spines. The name *quadracus* denotes the number of those spines. The first three are arranged in an alternating inclined left and right pattern. As in all sticklebacks, the last spine is attached to the first ray of the soft dorsal fin. In cross section this species appears triangular, and a lateral view reveals a very slender caudal peduncle. The body color is variably but usually olive green with dark blotches on the side and a silvery abdomen. During the breeding season, males are distinguished by bright red pelvic spines. The nest built by the male is cup shaped, not tunnel shaped, and often more than one nest is built one above another on an aquatic plant stem.

Threespine stickleback

Gasterosteus aculeatus

Hudson Bay, Canada to North Carolina
Max. length = 4 in.

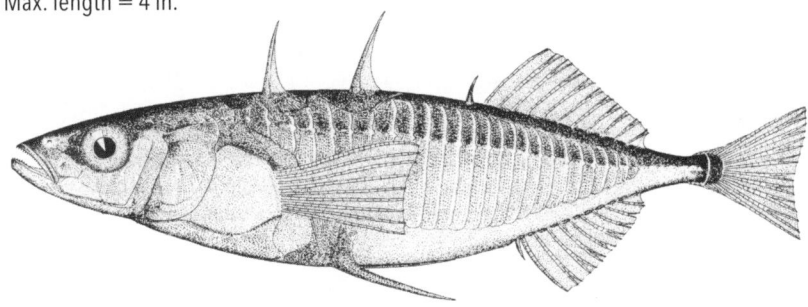

FIELD CHARACTERISTICS. *Gasterosteus aculelatus* (Greek terms meaning bony belly and spiny) refers to the ventral bony plate shielding the pelvic region and the first dorsal fin, which consists of two long free-standing spines followed by a shorter third spine that precedes the soft-rayed second dorsal fin. An even stouter spine constitutes each pelvic fin.

ECOLOGY/LIFE HISTORY. In the spring, males build nests made of green algae and filaments of other aquatic plants. The males begin by excavating a depression in the sandy bottom and gathering nest material that they mold and glue in place. The "glue" is a mucus produced by a specialized part of the male's kidney. The nest is complete when it is barrel shaped and has a distinct entrance and exit. This nest is commonly in the center of the male's territory. A larger, silvery, egg-bearing female is attracted to the nest by the zigzag mating dance of the brightly colored (red chest and blue eyes) male.

The female is led to the entrance of the nest and induced to enter and deposit a clutch of eggs. Simultaneous with her departure out the exit, the male enters and quickly fertilizes the eggs. The male then guards and takes care of these eggs. After hatching, the small fry swim out of the nest. During that time, the adult catches them in his mouth and spits them back into the nest. Soon, as the young fish grows, it learns to evade the parent and swims away as an independent fish. It has been suggested that this early experience training is later used by the young fish to better evade the attack of predatory fishes. As larger adults, the most conspicuous defense against predators is the sharp dorsal and pelvic spines. These spines are strong and can be locked into an erect position. Tests have shown that young 6 in. pike have difficulty swallowing adult threespine sticklebacks, and those predators soon learn to avoid sticklebacks as prey. Larger fish and fish-eating birds still are capable of preying upon sticklebacks. Because sticklebacks have small mouths, they tend to feed on small crustaceans and often fish eggs, even those of their own species.

The threespine stickleback is the most widely distributed stickleback species. It can be found on both sides of North America, in Europe, and in parts of Asia. Some populations are anadromous while others are strictly marine or freshwater. As a result, morphological variation is common and subspecies status has been given to some populations, in particular along the west coast of the United States and Canada. Threespine sticklebacks can be found in our bays, estuaries, and tidal creeks and marshes. Of all the stickleback species, *Gasterosteus aculeatus* became the most well-known because its territorial and reproductive behavior were studied by Niko Tinbergen who, later in 1973, won a Nobel Prize for his contributions to understanding animal behavior.

Blackspotted stickleback

Gasterosteus wheatlandi

Other common names: twospine stickleback
Newfoundland to Long Island
Max. length = 2 in.

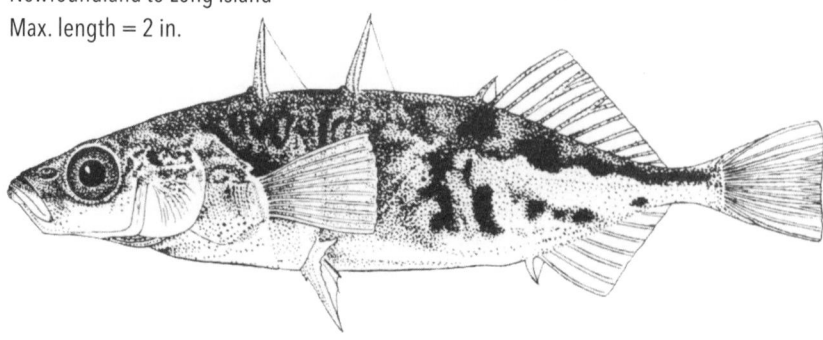

FIELD CHARACTERISTICS. The distribution of the blackspotted stickleback is the most limited of all our local sticklebacks. This species was named after R. J. Wheatland, who collected the first specimens at Nahant, Massachusetts, in 1859.

Even though this species also has three dorsal spines as in *G. aculeatus*, the first two spines are again the most prominent. This species can be distinguished from *G. aculeatus* by pelvic fin differences. In *G. aculeatus* the pelvic fin consists of one spine and no soft rays versus one spine and one–three soft rays in *G. wheatlandi*. Further, the base of the pelvic spine in *G. wheatlandi* has well-developed pointed cusps. Overall, the blackspotted stickleback appears slightly stouter than *G. aculeatus*, and in males during breeding the pelvic region is greenish yellow, brassy, or gold.

Ninespine stickleback

Pungitius pungitius

In eastern North America from the Arctic to New Jersey
Max. length = 3 in.

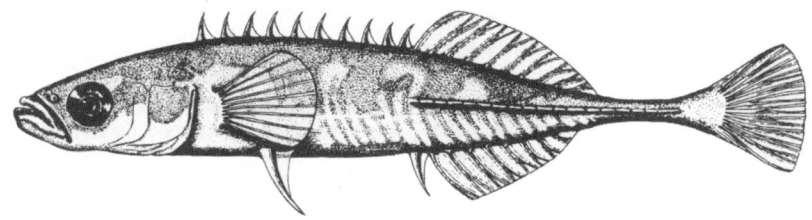

FIELD CHARACTERISTICS. *Pungitius* has a global distribution similar to that of *G. aculeatus*. It occurs on both coasts of North America as well as in Asia and Europe. Morphological variations in these widely spread populations have resulted in the recognition of several nominal species within this genus. In our waters, when it occurs, it is most commonly found in shallow, protected bays and coastal ponds.

In *Pungitius*, as well as other stickleback species, the number of spines in the first dorsal fin is commonly one particular number, but it is not absolutely fixed. Within the many populations of ninespine sticklebacks, the number of spines may range from 7 to 11. A common taxonomic method is to count features, such as scales, fin rays, spines, and so on. Those characteristics are called meristic traits. They are normally very reliable characteristics used for species identification. However, such traits tend to vary depending upon the prevailing conditions under which the particular fish develops. Factors that affect larval growth, such as temperature, salinity, or food availability, can alter the final number of a meristic character.

In addition to the number of dorsal spines, the caudal peduncle in *Pungitius* has a distinct lateral keel on each side. This feature is shared with most populations of *G. aculeatus* but does not occur in the two other local stickleback species. The body color is variable from olive brown to gray, but breeding males have been reported to have black abdomens.

LITERATURE. Hoogland et al. 1956; Reisman 1968; Wooton 1984; Bell and Foster 1994

PIPEFISHES AND SEAHORSES
Family Syngnathidae (*peces pipa y caballitos de mar*)

As different as pipefishes and seahorses might appear at first glance, they share enough basic characteristics as to form a single family. In some ways, seahorses can be considered as curious upright and curved pipefish. There are 51 genera and 190 species of pipefishes and only 1 seahorse genus containing 33 species. Within the family, these fishes are mostly found in marine and brackish waters, but a few pipefishes occur in freshwater. All the species have a tubular snout tipped with a small mouth formed from highly elongated bones of the anterior part of the skull (neurocranium) and the bones supporting the jaws (suspensorium). At short distances, small prey animals are quickly sucked into the toothless mouth. The small-sized prey includes copepods, amphipods, and fish eggs.

There are no pelvic fins and only one dorsal fin. The soft rays of that dorsal fin can undulate and contribute to forward motion. Most uniquely, males of all members of this family care for the eggs in a marsupial-like brood pouch after the eggs have been placed there by the female during mating. Hundreds of eggs may be transferred to the male over a series of repeated pairings with the same female. In that limited sense, they are monogamous, but courtship and mating need not involve the same pair during the entire breeding season.

The female's protruding oviduct is used to transfer eggs into the male's pouch and, at that moment, are fertilized. The eggs are embedded into the male's pouch, and there the embryos live off their yolk sac. Fluids in the pouch may also contribute to the nutrient and oxygen needs of the developing embryos before they are expelled as small, less than half an inch long, young fry. The incubation process usually varies from 10 to 14 days.

In Southeast Asia and Australia, "pipehorses" in the genus *Solegnathus* appear to be a transitional form of syngnathid. In that genus, the heads are angled slightly toward the trunks, and the elongated bodies have prehensile tails.

Lined seahorse

Hippocampus erectus (caballito estriado)

Nova Scotia to Uruguay
Max. length = 8 in., av. length = 4 in.

FIELD CHARACTERISTICS. Like pipefishes, seahorses have bodies encased in bony rings but in the seahorse those rings are more sharply pronounced and the tail in seahorses is prehensile. It uses that prehensile tail to hold on to vegetation or even other seahorses.

Undulations of the single dorsal fin and the paired pectoral fins allow seahorses to swim in a vertical position. The pectoral fins are placed just where the ears would be on a horse's head, adding to the very unfish-like appearance of seahorses. The genus *Hippocampus* is derived from the Greek for "horse sea monster." The lined seahorse is usually found in shallow waters but not always. The body color is highly variable and can be from light brown to yellow, green, or red. Sometimes, the top of the head is decorated by fleshy filaments whose function, other than to add to the fish's camouflage, is still undetermined.

ECOLOGY/LIFE HISTORY. In some countries, seahorses are collected, dried, and sold as traditional medicine. To prevent overharvesting and exploitation, all seahorse species are considered threatened and are protected under the Convention on International Trade in Endangered Species.

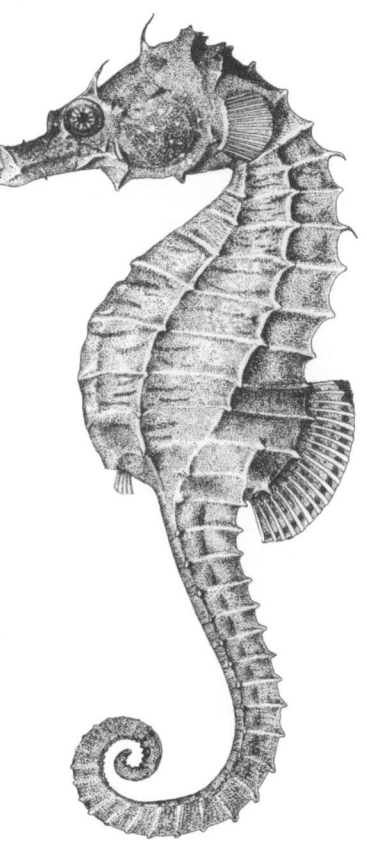

Northern pipefish

Syngnathus fuscus (pez pipa)

Gulf of St. Lawrence to Florida
Max. length = 12 in., av. = 8 in.

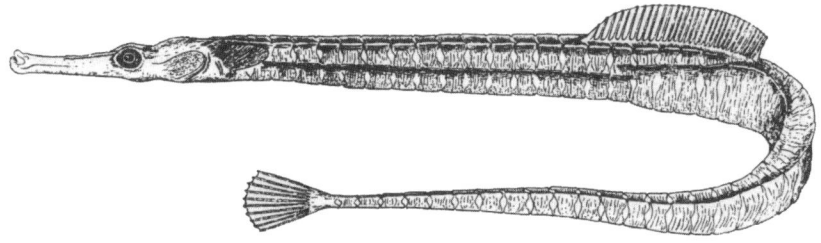

FIELD CHARACTERISTICS. Unlike seahorses, pipefishes have a rounded caudal fin and move horizontally by using the dorsal fin and eel-like movements of the tail. The body is mottled, and the color is highly variable, being greenish, olive, or brown. This is likely an adaptation to better match their habitat and be less conspicuous.

ECOLOGY/LIFE HISTORY. During mating, when mates court and intertwine, pipefishes can assume a seahorse-like upright position. The most common habitats are the shallow eelgrass beds or areas of dense seaweeds. Of the many species of pipefishes that occur along the Atlantic coast, the northern pipefish is the only one that occurs commonly north of New York to Canada. This species might be the most abundant and predictable fish found in our local, nearshore, eelgrass beds. In the winter, like many small, relatively slow-moving, resident fishes, this species moves to deeper waters where ice would not form.

LITERATURE. Bergert and Wainwright 1997; Wilson et al. 2003; Ripley and Foran 2009

CORNETFISHES
Family Fistulariidae (*cornetas*)

This family is widely distributed within the shallow coral reef habitats of tropical and subtropical regions of the Indian, Pacific, and Atlantic Oceans. There is only one genus, *Fisturlaria*, and it has four species.

Bluespotted cornetfish

Fistularia tabacaria (corneta azul)

Nova Scotia to Brazil
Max. length = 6 ft., excluding caudal filament

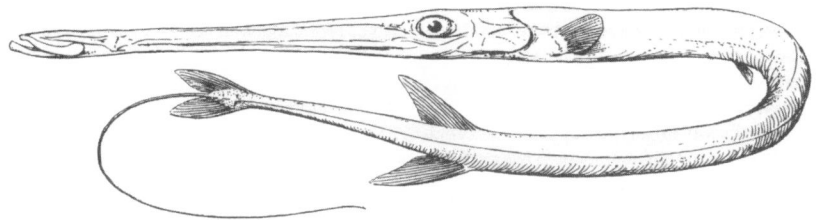

FIELD CHARACTERISTICS. All specimens observed in New York waters are juveniles and would be less than 1 ft. in length. This is an unusual fish any way one looks at it. The snout is almost one quarter the length of the fish's body, and a filament originating from the middle two rays of the caudal fin is about equal to the snout's length. The tubular snout is even longer than the one possessed by pipefishes, with which the cornetfish is often confused. Only the tip of the snout contains the bones of the upper and low jaws (premaxilla, maxilla, dentary, angular). The other parts of the snout are highly modified and extended elements of bones that generally support the jaws and articulate them to the skull. Bones that are part of the skull and the gill covers are also involved in the snout's structure.

The long tail filament in cornetfish might be a way to distract a fatal attack by a predator. Fish that might be prey to larger predators have a variety of ways to deter predators from identifying the sensitive head of a fleeing fish. For example, some fish have long vertical bars through the eyes or eyespots toward the back of the fish. The general olive-brown color of cornetfish can be influenced by the environment's background, but wherever they live, the distinctive and likely confusing blue spots on the snout and the body are always visible.

WARM-WATER EXPATRIATES

Locally the cornetfish can be found during the warm months within our bays and estuaries, but the most common habitat for this fish is the area on or around coral reefs. A significant number of marine fishes found in Long Island marine waters are more commonly found in warmer subtropical waters. These are not merely seasonal visitors who leave in the fall but are individuals who are transported as juveniles to the north during late spring and grow during the summer and early fall. When water temperatures decrease these fishes become moribund and die or are unable to escape local predators. Many of these subtropical expatriates are common members of fish communities typically living within turtle grass meadows, coral reefs, or on the sandy margins surrounding the reef. Some examples of the warm-water marine families that can be seen locally are goatfish, squirrel fish, bigeye, cornetfish, butterflyfish, angelfish, snapper, grouper, mojarra, jacks, surgeonfish, trunkfish, triggerfish, filefish, and porcupine fish. In general, these fishes do not spawn north of Cape Hatteras.

The primary mechanism for their transport to the north is as larvae or juveniles who are passively carried within the wide, warm waters of the Gulf Stream. This current originates in the Gulf of Mexico, exits and moves offshore between southern Florida and the Bahamas, and follows the East Coast northward. The Gulf Stream can be about 60 mi. wide and has a speed of several knots. Before it gets to the latitude of Long Island, the Gulf Stream turns to the northeast and is about 200–250 mi. south of Long Island. However, the current often meanders in such a fashion that it pinches off large eddies called rings. These rings can be over 100 mi. in diameter and can extend to great depths. Those rings that are formed on the north side of the current are called "warm core rings." These rings move westward along the continental shelf and toward the coast. (Warm core eddies at the edge of the continental shelf are often a place where tunas can be caught in the spring). Eventually the rings lose their structure, and some of those juvenile fishes, now moving along the Long Island coastline, will find themselves being entrained by flood tides through the available inlets and into the large south shore bays, that is to say, Shinnecock, Moriches, and Great South Bays. Once inside the bays, these fishes tend to seek refuge in the shallow eelgrass beds and over the summer and early fall tend to concentrate in number, and individuals grow in size. It is unlikely that they would be able to find the inlets and escape from the bays that now act as big fish traps.

LITERATURE. Hare et al. 2002; Wood et al. 2009; Able et al. 2013

SCORPIONFISHES
Family Scorpaenidae (*escorpiones*)

This is a very diverse family with 56 genera and 418 species and includes many of the most venomous marine fishes (e.g., scorpionfishes, rockfishes, stonefishes, and lionfishes). They are found in all temperate and tropical seas with most being native to the Indo-Pacific region.

Members of this family are characterized by prominent spines on their heads, gill covers (preopercular spines), and on dorsal, anal, and pelvic fins. Venom glands of scorpionfishes lie along glandular grooves in the upper half of the spines. When a spine punctures an object, the glandular tissue is ruptured and the venom is released into the wound.

Lionfish

Pterois volitans (pez leon rojo)

Other common names: turkeyfish, firefish, zebrafish
Indo-Pacific from the Red Sea to Pitcairn Island and invasive along
    the Atlantic coast
Max. length = 18 in.

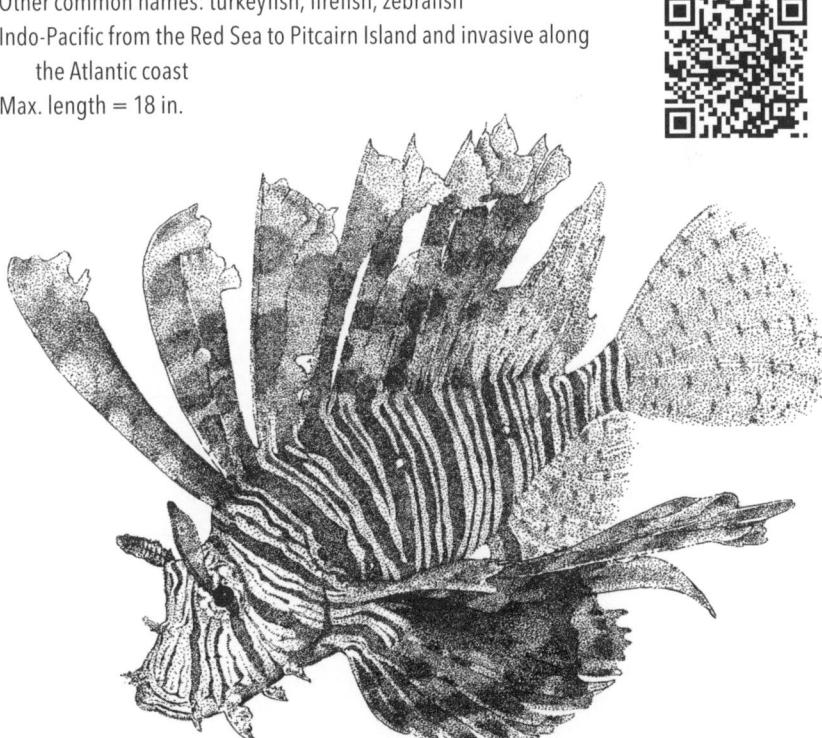

© Roger Hall Scientific Illustrations

FIELD CHARACTERISTICS. This species has colorful, enlarged, pectoral fins with 14 feather-like rays that inspired its scientific name, which means "winged" and "flying." Its 13 dorsal, 3 anal, and 12 pelvic spiny rays are equally striking but, unlike the pectoral fin rays, are venomous. It is suggested the garish barred, banded, and spotted brownish-red color pattern serves to advertise the fish's potent defenses, so that they are avoided by most potential predators. If the lionfish is threatened, it often assumes a head-down defensive position with dorsal spines erect. The neurotoxin in the venomous spines can cause severe pain in humans and has led to fatalities in some cases.

ECOLOGY/LIFE HISTORY. The lionfish is one of the few Pacific marine fish species that was released into the Atlantic to have subsequently dispersed and established invasive breeding populations. One suspicion is that the lionfish was introduced during Hurricane Andrew in 1992, presumably as an accidental release of aquarium fish. Lionfish do not have many natural predators and do not usually occur in high densities in their native Indo-Pacific habitats. However, since their introduction to the warm waters of the western Atlantic Ocean, their populations have exploded, and this species has become a menace to native reef fish communities in the Bahamas, Belize, and other parts of the Caribbean. One study of a Bahamian reef documented a five times denser lionfish population than a comparable Pacific habitat. The few reports of Indo-Pacific lionfish predators include those involving cornetfishes, some groupers, and scorpionfishes. As lionfish become more numerous in the warm western Atlantic, spear fish and commercial fishers may be their most formidable predators. Also, fishing contests and tournaments in many Caribbean Islands and Florida are used to help keep populations somewhat under control.

The introduction of the lionfish was either intentional or accidental but because of its fecundity, long reproductive season, and lack of predators, this inoculation of lionfish spread quickly to the north and south. At present, there are reproductive populations of lionfish in North Carolina and within the Caribbean region. In September 2001, two juvenile lionfish were collected along the south shore of Long Island. Now, lionfish can be occasionally observed with other tropical fishes that are swept northward to our shores in the summer and fall. None of these warm-water visitors, including the lionfish, can survive our icy winters.

LITERATURE. Whitfield et al. 2002

SEAROBINS
Family Triglidae (*vacas*)

This is a benthic family with 10 genera and about 105 species. The most distinguishing characteristics are the large and expansive pectoral fins with the lower three rays being free and able to move independently. As their common name might suggest, searobins are sound producers.

Northern searobin

Prionotus carolinus

Other common names: Robin
Bay of Fundy to Florida
Max. length = 17 in.

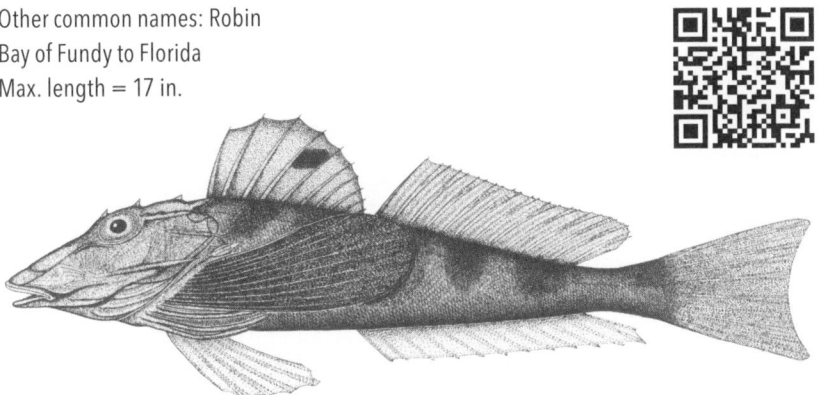

FIELD CHARACTERISTICS. This species is very similar to *P. evolans* but has no longitudinal stripes and has shorter pectoral fins. The northern searobin's pectorals end between the origin and center of the anal fin base whereas the striped searobin has a pectoral fin terminating much closer to the end of the anal fin.

ECOLOGY/LIFE HISTORY. Searobins live on the bottom and use the thick, flexible three lower rays of the pectoral fins to probe the muddy and sandy substrate for food. Embedded within the surface of these specialized fins are millions of solitary chemosensory cells that serve like taste buds and detect food. Among the prey are shrimp, crabs, clams, and marine worms.

Searobins make sounds when disturbed and are most often heard during the summer breeding season. As in some other sound-producing fishes, the sounds are made by well-developed intrinsic sonic muscles incorporated into the wall of the swimbladder. These muscles drum against both sides of the gas-filled swimbladder. The large bony plates and spines on the head and the 10 strong spiny rays of the first dorsal fin may help deter all but the largest predators, such as sharks, bluefish, and goosefishes.

LITERATURE. Kotrschal 1996; Connaughton 2004

Striped searobin

Prionotus evolans

Bay of Fundy to northern Florida
Max. length = 19 in., av. = 10 inches

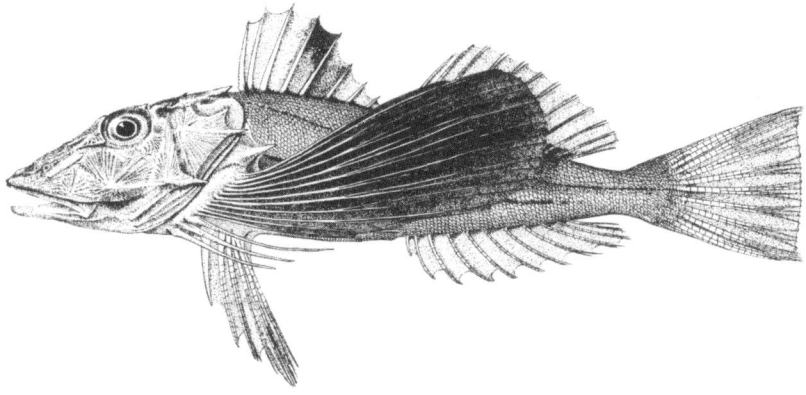

FIELD CHARACTERISTICS. The striped searobin is distinctly marked and colorful. As the common name would lead one to expect, there is a stripe along the lateral line and one more located ventrally. The body is pale brown, but the pectoral fins are brownish orange. The large pectoral fins of the searobin may look like those of a flyingfish, but this fish does not take flight.

SCULPINS
Family Cottidae (*charrascos espinosos*)

This family has about 70 genera and 275 species. Most species are coastal marine and are distributed throughout the Northern Hemisphere. Some are freshwater and there are a few deep-water marine species in the vicinity of Australia and New Zealand. All three relatively temperate, nonpolar, inshore Atlantic species are found in New York waters.

Sculpins possess large, fan-like pectoral fins. The head is somewhat broad and depressed with ridges and conspicuous spines. The relatively large eyes are placed high, and the space between the eyes is narrow. The gill rakers are short, suggesting these fishes tend to feed on a range of small marine organisms rather than plankton.

Grubby

Myoxocephalus aenaeus

Other common names: little sculpin
Newfoundland to New Jersey
Max. length = 7.5 in., av. length = 5 in.

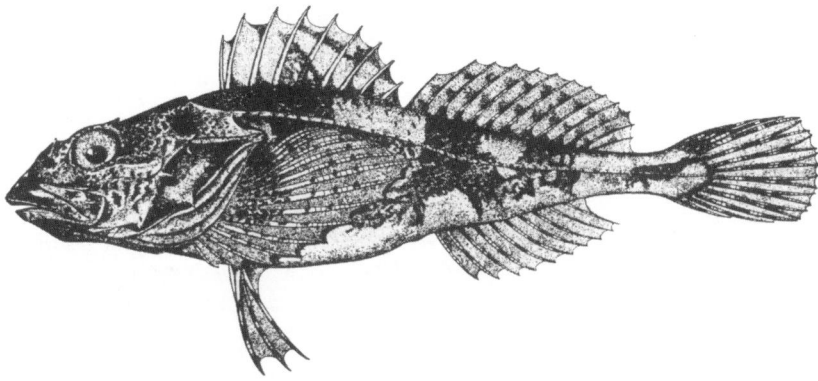

FIELD CHARACTERISTICS. The bottom-dwelling grubby has a distinctive mottled color pattern that adapts well to the irregular features of the substrate. The grubby is the most common shallow-water sculpin and is found on diverse substrates and not infrequently near eelgrass beds. The other New York sculpins are more likely found in higher saline, deeper, and colder waters.

This species can be collected within the bays and coastline of New York throughout the year. Spawning by the grubby occurs from December to March. Their shallow-water, benthic, spawning sites are occasionally covered by subzero ice-laden waters. In order to avoid freezing, these and other local fishes (e.g., winter flounder and tomcod) have evolved the ability to seasonally synthesize antifreeze molecules. This smallest of the local sculpins has no commercial value although the other local sculpins may have been used historically as lobster bait.

Longhorn sculpin

Myoxocephalus octodecemspinosus

Other common names: hacklehead, gray sculpin
Newfoundland to Virginia
Max. length = 18 in.

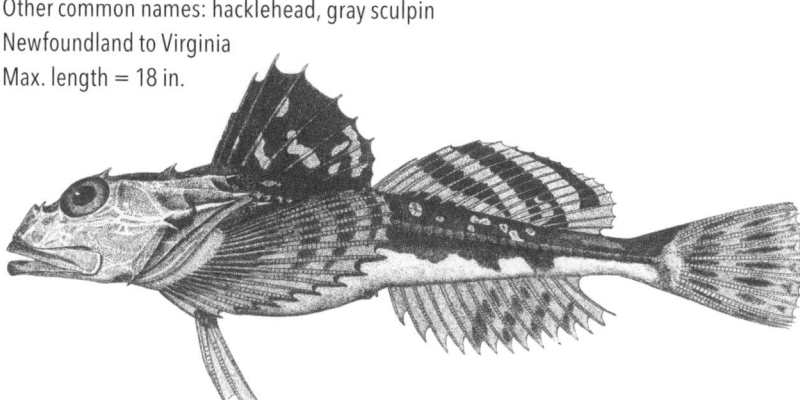

FIELD CHARACTERISTICS. *Octodecemspinosus* refers to 18 spines in the head region; although that is a formidable number, the count, according to Jordan and Evermann, is actually 20, which would have created the equally unpronounceable species name of *vigintispinosus*. The longest cheek spine reaches back to the margin of the gill cover, in contrast to the shorthorn sculpin whose longest cheek spine reaches only as far as half that distance.

ECOLOGY/LIFE HISTORY. The longhorn sculpin occasionally produces dull growling sounds. Unlike most fishes that produce sound, this species does not use vibrating muscles associated with the swimbladder. Sculpins have no swimbladder. In the longhorn sculpin and probably in the other local sculpins, sound results from the effect pectoral muscles have on pectoral girdle movements.

Shorthorn sculpin

Myoxocephalus scorpius

Greenland and Labrador south to New Jersey, northwestern Atlantic, and Arctic region.
Max. length = 19 in., av. =11 in.

FIELD CHARACTERISTICS. This species differs from the longhorn sculpin in having a shorter upper cheek spine.

LITERATURE. Fish and Mowbray 1970

SEA RAVENS
Family Hemitripteridae (*charrascos Cuervo*)

This family has three genera and eight species but only one species is found along the Atlantic coast. Sea ravens were formerly placed within the sculpin family. Both have two dorsal fins, but sea ravens have a first dorsal fin that is deeply notched between the spines. This distinction is why the scientific name *hemitripterus* (suggesting the presence of almost three fins) was used.

Sea raven

Hemitripterus americanus

Other common names: red sculpin, sea sculpin, raven
Labrador to Chesapeake Bay
Max. length = 25 in., av. =18 in.

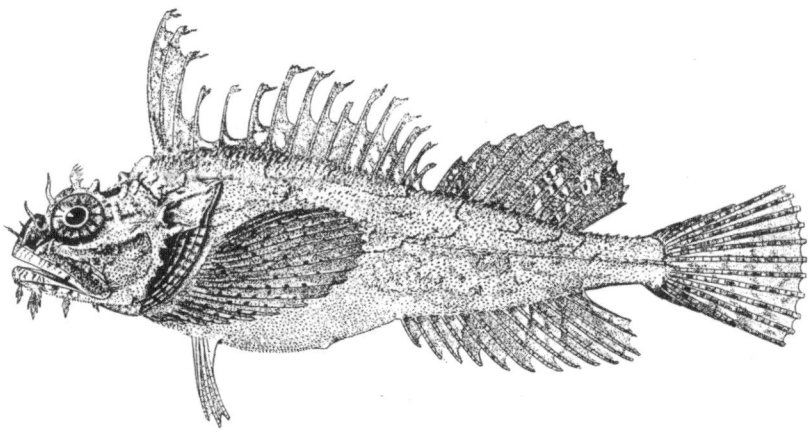

FIELD CHARACTERISTICS. This is an unusual looking fish. The upper part of the head has irregular features, and the lower jaw and head have fleshy flaps obscuring the large mouth and eyes. Skin flaps also occur at the ends of the spiny rays in the first dorsal fin. The body's surface is prickly, and the color is variable but can be reddish to brown whereas the belly is often yellow.

ECOLOGY/LIFE HISTORY. From the body shape, decorative body features, and rows of sharp teeth, it is clear that this species is benthic and carnivorous. It feeds on a variety of invertebrates and fishes. The reddish color might also suggest that it tends to live in deeper water rather than in shallow estuaries. Long wavelengths (red) are absorbed first and penetrate only to a depth of 50 ft., so a red-colored fish is essentially black below that depth and thus is difficult to detect.

In general, this is a cold-water species and shares the ability to produce antifreeze proteins with other fishes that are threatened by freezing temperatures. Along our coasts, the sea raven tends to move farther offshore during warmer months of the year.

LUMPFISHES
Family Cyclopteridae (*peces grumo*)

There are six genera and 28 species in this family that generally are found in the cool, marine waters of the Northern Hemisphere.

Lumpfish

Cyclopterus lumpus

Other common names: lump sucker
Greenland to Chesapeake Bay and the eastern North Atlantic
Max. length = 23 in., av. = 14 in.

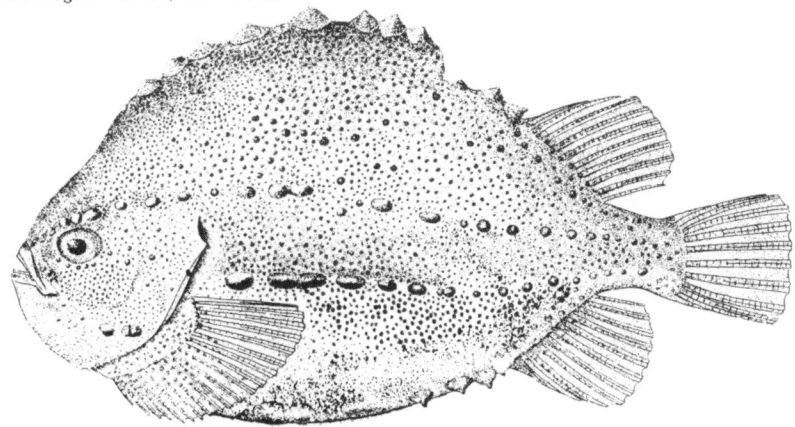

FIELD CHARACTERISTICS. The name *lumpus* describes this fish well. This species has a globiform body with the body depth measuring one-half of the fish's total length. The body is scaleless but is covered by small, irregularly spaced tubercles as well as a series of larger pointed tubercles arranged along the sides of the body. The fish has two dorsal fins but the first is not easily recognizable as containing spiny fin rays. Underneath the dorsal fins is an extensive mass of blue gelatinous cartilage.

The fish is often found holding fast to a bottom structure. This is made possible by the pelvic fins modified into a single large ventral suction device. This is comparable to the pelvic fin modifications found in fishes belonging to other families (e.g., gobies, clingfishes, and freshwater algae eaters).

The commercial value of this species is from harvesting the roe from fecund (often more than 200,000 eggs) females and selling that as affordable caviar.

LITERATURE. Budney and Hall 2010

TEMPERATE BASSES
Family Moronidae (*lobinas nortenas*)

There was a time when these basses were included within the sea bass family, Serranidae, but there are some distinctions that separate the two groups. Moronid species have two, rather than three, spines on the posterior margin of the gill cover. The lateral line extends on to the caudal fin, but in serranids the lateral line ends at the caudal peduncle and does not extend farther.

This family is native to North America although some of its species have been introduced into Europe and North Africa.

Morone is the most prominent genus within the family. This genus has four species in North America, two of which, the yellow bass (*M. mississippiensis*) and the white bass (*M. chrysops*) are confined to freshwaters. The white bass is found in many of New York's lakes.

White perch

Morone americana

Gulf of St. Lawrence to South Carolina, most common from Hudson River to Chesapeake Bay
Max. length = 19 in., common = 9 in.

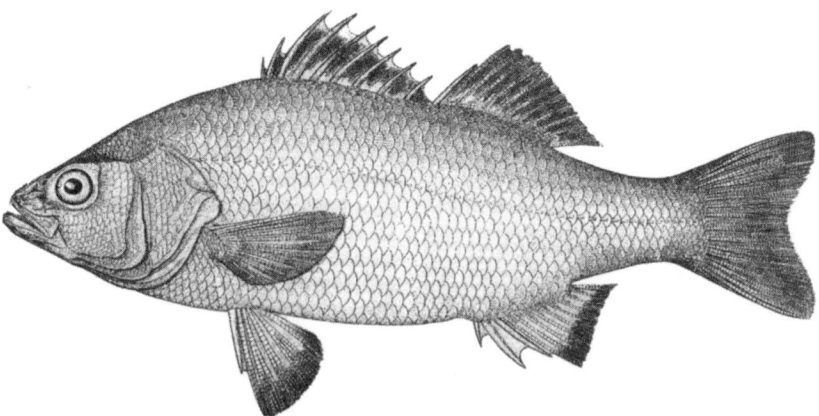

FIELD CHARACTERISTICS. In contrast to its larger and more sought after striped relative, the dorsal fins of the white perch are joined at their bases whereas those fins are separated in the striped bass. Juvenile white perch have pale stripes but they fade with growth.

ECOLOGY/LIFE HISTORY. White perch are euryhaline, that is, can be found in a wide range of habitats from freshwater to coastal marine, and thus have a relatively diverse diet of worms, shrimp, and small fishes.

Striped bass

Morone saxatilis (lobina estriada)

Other common names: rockfish, striper, roller
Gulf of St. Lawrence to Louisiana and introduced on the Pacific coast.
Max. length = 6 ft., max. weight = 125 lb., av. weight = 5–10 lb.

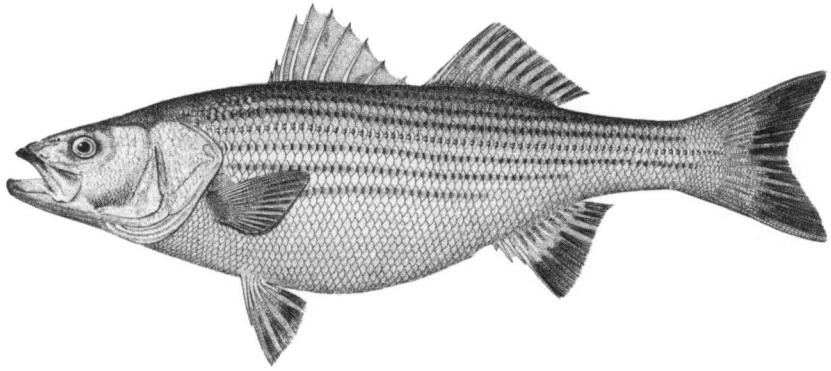

FIELD CHARACTERISTICS. As the name suggests, there are seven–eight conspicuous horizontal stripes distributed above, below, and along the lateral line. The teeth are not sharp or designed for cutting but are employed to grasp prey.

ECOLOGY/LIFE HISTORY. This is one of the best-studied anadromous fish. The typical striped bass spends the winter in deep Mid-Atlantic coast waters. As these waters warm, the fish migrates north and may detour to spawn. Spawning occurs within four major river systems: the Chesapeake, Hudson, Roanoke, and Delaware. Spawning takes place in freshwater often near the interface between the freshwater and the tidal salt front. Juveniles may overwinter near the estuaries entrance or in coastal marshes. After a year or two, the young move out to sea and return after 4 to up to 7 years to their natal rivers. Chesapeake adults spawn beginning in March whereas spawners move into the Hudson in April and continue to do so until mid-June. After spawning some may migrate farther north and stay there until the fall cooler weather, when they return to southern waters. The major source of stripers that run along our shoreline is the Chesapeake, which contributes to 90% of the total catch both commercially and recreationally. Although the Hudson stock is a distant second, one might expect that stock to be better represented in the Long Island Sound.

As in many fishes, females grow larger than males. Indeed, any striper over 30 lb. is very likely a female "cow." Large females are at a premium because of their capacity to produce numbers of eggs. A 5 lb. female may lay a half million eggs, and a 50 lb. female may lay five million.

Adults are voracious, opportunistic feeders, feeding on many smaller fish species and a variety of invertebrates.

LITERATURE. Waldman et al. 1988

FISHERIES. Of all the nearshore coastal fishes in southern New England and the Mid-Atlantic, striped bass is at the top of the list of highly valued catch for both recreational and commercial fishermen. No other inshore fish generates as much interest, excitement, emotion, and debate as does striped bass. For commercial fishermen, striped bass is a "money fish" as they typically return a very high price per pound to fishermen. Of course, this high market price is driven by high market demand as they are a favorite of consumers and appear on many white tablecloth restaurants during season. Striped bass is the most prized catch of the nearshore angler both by boat and from shore (surf casting). Striped bass are fun to catch as they fight hard and long, and catches of fish averaging 20 to 30 lb. are not uncommon and fish into the 50 lb. class can occur. Striped bass can at times be very stealthy, picky, and difficult to catch; while at other times they congregate in large schools or "blitzes," where fishermen can catch them one after another. The fall run of striped bass along Long Island is legendary, with Montauk being "the Mecca" for striped bass anglers. Striped bass has a special place in the hearts of recreational fishermen. Striped bass can test the ability of fishermen, and individual anglers are often judged based on their success, or lack thereof, of striped bass catches in terms of numbers and size of individual fish.

Unlike many other important nearshore species, maximum striped bass commercial harvest took place in the 1960s and 1970s rather than the early part of the century. Peak commercial landings in New York occurred in 1973, when 1.7 million pounds were landed. Landings fell considerably during the 1980s as abundance of the resource declined. Landings also declined because of management moratoria and restrictive regulations. The commercial fishery in the Hudson River closed in the 1970s because of PCBs in the fish and has not reopened. The commercial fishery is regulated by quota, and New York landings averaged 700,000 to 800,000 lb. from 2003 through 2017. The 2019 determination of overfishing (see below) resulted in a coast-wide harvest reduction of 18%. New York landings since then have been around

575,000 lb. Since 2020 coast-wide commercial landings have been around 4 million pounds.

Although not totally closed during 1986–1989 as was the commercial fishery (and the recreational fisheries in some other states), severe catch restrictions limited the recreational fishery during that rebuilding period. Since 2003 the East Coast recreational fishery has harvested an average of 49 million pounds annually. During the same time period the New York recreational harvest has averaged 11 million pounds. Landings have recently been reduced because of reduced abundance of striped bass and a reduction in the allowable recreational harvest amount as of 2020 (see Management below) and landings have averaged around 1.75 million pounds annually in New York. However, a significant increase in New York recreational landings in 2022 to 10.7 million pounds along with increases in other states prompted ASMFC in 2023 to implement emergency measures (see Management below). New York anglers catch around 24% of the coast-wide recreational catch. Discard mortality in the recreational striped bass fishery—the number of fish that die after being released—has become problematic and has contributed to overfishing of striped bass. The latest stock assessment (NMFS 2019) indicates that recreational discard mortality now exceeds the recreational landings. The assessment uses a 9% discard mortality rate (9% of fish released will die). Improved handling of striped bass and the use of circle hooks can help to reduce this mortality (https://ccesuffolk.org/marine/fisheries/recreational-striped-bass-outreach-and-education).

ALL TACKLE WORLD RECORD: 81 lb. 14 oz., Long Island Sound, Connecticut

NEW YORK RECORD: 76 lb. 0 oz.

MANAGEMENT. Striped bass is managed by the ASMFC and the NYSDEC through the Interstate FMP for Atlantic Striped Bass (ASMFC 1981b) and the plan's subsequent amendments and addenda. In 1984 the U.S. Congress passed The Atlantic Striped Bass Conservation Act, which provided the coastal states the necessary tools and enforcement ability to cooperatively and effectively manage striped bass stocks. The act also provided for a mechanism for the federal management of striped bass in waters outside of 3 mi. Unlike many other species, striped bass is not managed federally through the Fishery Management Council process. It is currently illegal to harvest striped bass by any means outside 3 mi.

Owing to heavy commercial and recreational fishing pressure and environmental issues, the striped bass population experienced overfishing and a significant decline in population size during the 1980s. A moratorium was imposed in the commercial fishery from 1986 through 1989, and severe restrictions were placed on the recreational fishery. As the resource rebuilt, the commercial fishery was re-opened in 1990 and recreational restrictions were liberalized. The resource was declared rebuilt in 1995. Striped bass is currently sustainably managed. However, owing to the strong influence of environmental conditions during spawning that affect year-class strength as well as year-to-year variability of fishing mortality, the overfished/overfishing status of the resource changes with updated stock assessments. However, the FMP mandates that any overfished/overfishing determinations be immediately addressed. As of the publication date of this book, striped bass is again in a rebuilding mode due to their overfished status (but overfishing is not occurring) (NMFS 2019). Management measures initiated in 2020 included an 18% commercial and recreational harvest reduction. Regulatory measures currently include a recreational slot size (a minimum and a maximum size) and a commercial minimum size limit (in New York also a commercial slot size), recreational bag limits, seasons, and commercial quotas. The basis of the management program is to keep the spawning stock biomass at or above its 1995 level.

In New York commercial fishermen are each issued single-use tags that have to be affixed to the jaw of every fish sold. The tags are issued annually, and each recipient's number of tags is based on New York's quota, the total number of licenses, and full-time or part-time qualification. Participation in the fishery is currently limited to only qualified fishermen with a striped bass catch history. Historically most commercial striped bass were caught by haul seines until it was ruled in New York to be an illegal gear for the harvest of striped bass as a measure to reduce discard mortality and to reduce user conflicts on the beaches. Striped bass are currently caught commercially by gillnet, pound net, trawl, and hook and line. New York commercial fishermen harvest around 30% of the coastal quota, second only to Massachusetts (the Chesapeake Bay has a separate quota).

In 2023 the ASMFC took emergency measures to further restrict the recreational catch and initiated a new addendum to help rebuild the stock by 2029.

SEA BASSES
Family Serranidae (*serranos*)

There are, in total, 50 species of sea basses that occur in the Atlantic waters ranging from Canada to Mexico. A few have common names that include the word "bass," which tends to serve the confusion some have with the unrelated freshwater basses and striped bass. Many of the 475 species that are classified as serranids worldwide are the better-known tropical species such as groupers and hinds, among others. This is a very diverse family in terms of size and habit. In warmer waters along the Atlantic coast, these species range from small colorful schooling planktivorous reef fishes (the anthiines), to the territorial 8 ft. Goliath grouper that can exceed 800 lb. A fish of that size can eat anything it wants from spiny lobsters to stingrays to turtles.

The gill cover in this family has three spines and the dorsal fin is long and singular. In contrast, the temperate bass family, Moronidae, has only two opercle spines and the dorsal fin is in two distinct sections.

Many species of serranids are protogynous hermaphrodites who are initially female and then transform into males.

Black sea bass

Centropristis striata

Nova Scotia to Florida
Max. length = 2 ft.

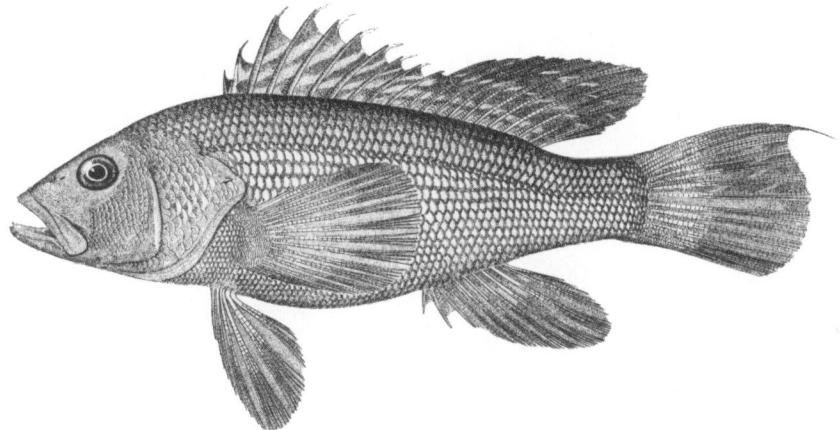

FIELD CHARACTERISTICS. The overwhelming appearance is a gray body with areas that can be intensely black and scales with small white centers. There are some lighter areas interrupted by broad bars and sides with dark stripes, thus justifying the species name of *striata*. The dorsal and anal fins have white trim and the caudal fin edge has white filaments with the upper caudal ray being somewhat longer.

ECOLOGY/LIFE HISTORY. Hermaphroditic females start to show hints of transitioning to males at about 3 years of age. In general, the oldest female is about 5 years old. Mature females tend to have a yellow-green hue body color.

Males are more than 6 years old. Dominant adult males develop a bluish hump behind their head during the spawning season.

Knowing the age structure is important to the sustainable management of a population of fish. Fishes are aged by microscopic examination of various hard parts including scales, otoliths (ear bones), spines, and vertebrae. These structures have growth rings similar to a cross section of a tree trunk and can thus be used to age the fish.

The black sea bass is omnivorous but prefers crustaceans and bivalves.

LITERATURE. Lavende 1949; Cochran and Grier 1991

FISHERIES. Black sea bass is an important and popular component of the catch of both commercial and recreational fishermen throughout the Mid-Atlantic and into southern New England. Sea bass is a very flavorful, if boney, fish that is often served whole (scaled and gutted) and has high market demand. Sea bass is often held and marketed live to promote absolute freshness and flavor and increased value. Sea bass is prized by recreational fishermen also for their flavor and for their feistiness when they bite the hook. Sea bass tend to congregate on wrecks, rocky bottoms, and jetties and can often be caught one after another. Fisheries change seasonally in relation to sea bass migration and distribution. They are caught inshore during warmer months and offshore during winter.

Like many other nearshore species, sea bass became increasingly important in New York commercial landings during the early and middle twentieth century. New York commercial landings reached a peak of over 2.5 million pounds in the early 1950s. The coast-wide commercial landings for the Mid-Atlantic/southern New England also peaked in the early 1950s at around 22 million pounds. Landings have dropped off considerably since then and averaged 220,000 to 300,000 lb. in New York during the last part of the twentieth century. Landings have been regulated by quota since 1998. However, as biomass has increased and the New York percent allocation of the total quota has increased, New York landings since 2020 have been 400,000 to 450,000 lb. Otter trawls and fish pots have accounted for most of the landings in New York. Other important commercial gear in New York includes pound net, hand lines (rod and reel), and lobster pots.

Sea bass is highly sought by recreational fishermen in New York and throughout the Mid-Atlantic. Most activity takes place inshore during spring, summer, and autumn. The Atlantic coast sea bass recreational fishery from Cape Hatteras and north peaked in 2016 at 12.8 million pounds. In New York, recreational catches have been highest since 2000, peaking at 6.5 million pounds in 2016, reflecting an increase in abundance as well as a northern shift in the distribution of sea bass. New York recreational landings have averaged just over three million pounds since 2016.

ALL TACKLE WORLD RECORD: 10 lb. 4 oz., Virginia

NEW YORK RECORD: 9 lb. 0 oz.

MANAGEMENT. Black sea bass is managed sustainably by the MAFMC, the ASMFC, NMFS, and the NYSDEC through the Summer Flounder, Scup and Black Sea Bass FMP (MAFMC, 1987). The management program currently allocates 55% of the total annual quota to the recreational fishery and 45% to the commercial fishery. As of the publication date of this book, black sea bass is not overfished and overfishing is not occurring (NMFS 2020).

The coast-wide commercial quota is further divided into state-by-state quotas with coast-wide management measures including minimum fish and mesh sizes as well as pot/trap specifications. New York further imposes commercial trip limits. State-by-state commercial quotas are also modified by a regional biomass distribution factor to reflect stock status and the northern expansion of the range of sea bass. As of the publication of this book, New York is allocated 9.79% of the coast-wide commercial quota. The recreational fishery is managed through a recreational harvest limit, possession limits, seasons, and minimum sizes. The recreational sea bass fishery is further subdivided into regional management areas to provide flexibility along the coast while at the same time providing consistency of regulations among the states included in each region. New York is included with the states from Massachusetts to New Jersey in the recreational regional approach. Regional recreational regulations only apply to state water (within 3 mi.). In federal waters (outside 3 mi.) the regulations are consistent from Maine to North Carolina.

TILEFISHES
Family Malacanthidae (*blanquillos*)

This family includes 40 species with representatives found in all the world's major oceans.

The dorsal and anal fins of tilefishes are long and continuous and include spiny rays that precede more numerous soft rays.

As striking as those long unpaired dorsal and anal fins are, the condition of the pelvic and pectoral fins is typical of the diverse families that belong to the large fish order of Perciformes. This order includes 160 families (ranging from the temperate basses to the butterfishes) with approximately 10,000 species. All those species possess pelvic fins that have one spine and five soft rays and a pectoral fin that is placed laterally and immediately above the thoracic pelvic fins.

Tilefish

Lopholatilus chamaeleonticeps (conejo amarillo)

Nova Scotia to Gulf of Mexico
Max. length = 3.7 ft.

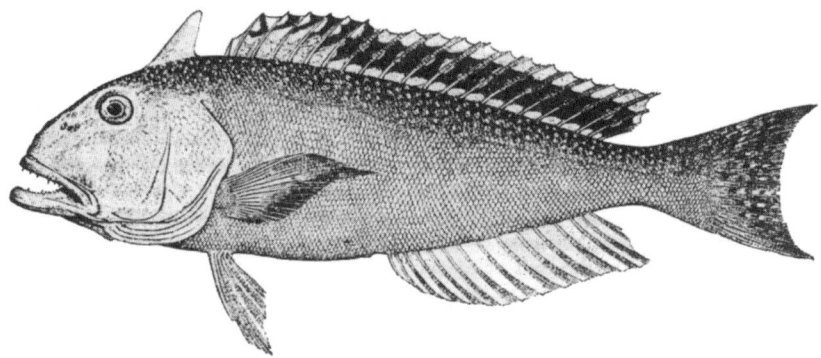

FIELD CHARACTERISTICS. This is the largest and the most colorful of the six
tilefish species that occur along the Atlantic coast. The dorsal surface from its
large convex head to its lunate caudal fin is bluish to olive green, including
an extensive field of small yellow spots on the side. The ventral area is
generally white, and shades of pink and purple might be found on the fins.
Although rapid color change is not dramatic, the "chameleon head" meaning
of the species name seems appropriate. Further, this species has a distinctive
predorsal yellow adipose flap on its head.

ECOLOGY/LIFE HISTORY. This is a benthic deep-water species that prefers
temperatures ranging from 47 to 53 °F (9–14 °C). Thus, the depth at which it
is found varies along its north–south range from 240 to 1400 ft. These depths
generally occur on the upper slope of the outer continental slope. Tilefish are
nonschooling, territorial fish. Individuals make a burrow in a relatively hard
sandy bottom or adopt a pre-existing crevice. There may be many thousands
of these "homes" per square kilometer. Canine teeth serve to crush the
mollusks and crustaceans that dominate its diet.

LITERATURE. Grimes et al. 1986

FISHERIES. Although there was a substantial commercial tilefish fishery in New
York in the 1930s, the fishery dropped off after that and only small amounts
were landed. There was resurgence in the commercial fishery in the 1970s as a
directed longline fishery for tilefish expanded rapidly in New York and New
Jersey. Peak landings occurred in New York in 1987 at 4.4 million pounds.
Landings dropped off in the 1990s because of reduced abundance and the

implementation of management measures. The fishery has experienced resurgence as the resource has rebuilt, and commercial landings since 2000 have averaged between 1.0 and 1.5 million pounds and are controlled by a quota. The fishery is conducted primarily by bottom longline gear with some incidental catch by trawls. The fishery takes place exclusively offshore as the fish do not migrate. Tilefish has high domestic market demand and thus retain a high price to fishermen. Tilefish is among the top five finfish in value landed in New York. Tilefish landed in New York is often referred to as golden tilefish. There is also an emerging blueline tilefish fishery developing in states south of New York, but that fishery has yet to become important in New York.

Tilefish is also a popular recreational species. However, because they are caught only offshore, angler participation and thus total recreational catches for New York are low. The low participation rate results in the recreational fishery being poorly accounted for in recreational catch surveys.

ALL TACKLE WORLD RECORD: 65 lb. 3 oz., New Jersey

NEW YORK RECORD: 63 lb. 8 oz.

MANAGEMENT. Tilefish is managed sustainably by the MAFMC and NMFS through the Tilefish FMP (MAFMC 2000). The tilefish stock has been rebuilt. As of the publication date of this book, it is not overfished and overfishing is not occurring (NMFS 2022b). The commercial fishery is heavily managed with limited entry, individual fishing quotas in the directed fishery, and an overall quota. There is also a prohibition on discarding in the directed fishery and no minimum size. There are trip limits in the incidental fishery. Recreational management measures include an overall recreational harvest limit and a bag limit. New York State does not have any management measures for tilefish.

BLUEFISH
Family Pomatomidae (*anjovas*)

There is only one species in this family but because of its worldwide distribution in temperate and subtropical waters, other than the eastern Pacific, there are many distinct breeding stocks. Along the East Coast of the United States there are six such stocks.

Bluefish

Pomatomus saltatrix (anjova)

Nova Scotia to South America (in the western Atlantic)
Max. length = 3.8 ft.

FIELD CHARACTERISTICS. Adults are generally blue green above and silver below whereas juvenile snapper blues have silvery sides. The body shape is one that is typical of strong, fast-swimming fish. The spiny portion of the dorsal fin and the pectoral fins fold into grooves that help reduce drag while swimming rapidly. The lower jaw is slightly extended and like the upper jaw bears uniformly razor-sharp and conical teeth.

ECOLOGY/LIFE HISTORY. The bluefish is perhaps the most frequently caught fish by recreational anglers. With their sharp teeth and strong jaws, freshly caught bluefish need to be handled with care. Their voracity is legendary as attested to by Spencer F. Baird, who in 1871 became the first commissioner of fish and fisheries of the U.S. Fish Commission, which is now known as NOAA's National Marine Fisheries Service. Baird likened the bluefish "to an animated chopping machine, the business of which is to cut to pieces and otherwise destroy as many fish as possible in a given space of time" (Greenwood and Norman 1975). He further goes on to describe the common observation that if the prey is too large to swallow whole, the hind portion is bitten off, leaving the head to sink uneaten. Swimmers and waders should be cautious during those times when large schools of bluefish may be chasing prey (e.g., menhaden) into the surf zone. It has been estimated that large schools of adult bluefish can consume hundreds of millions of fish during the summer season. At least 70 prey species have been found in stomachs of adult bluefish. Only the most inaccessible of fishes and invertebrates escape being in the adult bluefish diet. Juvenile snapper blues likely feed on a large number of silversides, mummichogs, anchovies, clupeids, and sand lances. Whenever bluefish arrive in our bays and harbors, they effectively reduce the forage fish populations. As an example, bluefish and the common tern have a commensal and competitive relationship. During the summer, feeding activity of blues drive prey to the surface, concentrating them, and facilitating capture by diving terns, although the benefits are short-term considering the large number of blues feeding on those fishes, thus depressing the resource for the terns.

Predators of adult bluefish must be large and swift, such as sharks and billfishes. Bluefish have an average swimming speed of about 3.5 mph but have been timed to swim more than twice that speed when startled. The latter ability gives them some chance to outmaneuver slower predators and chase down prey.

At speeds greater than 1.1 body lengths/second, bluefish ventilate their gills passively via highly efficient ram-gill ventilation. In fishes, conventional gill ventilation is accomplished by synchronous expansion and contraction of the buccal and opercular cavities to provide a nearly continuous, unidirectional flow of water whereby oxygen always diffuses into the gills. Because blood in the gills flows counter current to the incoming water, there is always a favorable exchange gradient whereby oxygen in the water always diffuses into the gill's blood. At high swimming speeds fishes such as bluefish and striped bass switch to ram-gill ventilation. This type of ventilation is common in tunas and sharks.

Bluefish is a migratory oceanic species that move into New York waters from mid-spring through autumn. Bluefish generally school by size, with school numbers that can at times be very large. In general terms, bluefish migrate northward and inshore in the spring and summer, and southward and offshore to the continental shelf in the fall and winter. These movements correspond to critical water temperatures (upper 50s Fahrenheit) and the movement and growth of their primary prey species.

Bluefish attain a large size at the end of their first growing season. They can grow 8 in. by the end of that first summer. They can do so because they have a size advantage over their principal prey. This advantage occurs because the young bluefish we see in our estuaries in June and July were spawned earlier (March and April) in southern offshore waters (the South Atlantic Bight) and had been advected to our latitude by way of the Gulf Stream. Those bluefish thus get a growth head start and gain a predatory advantage over their most common prey (silversides). A second, more modest, spawning event in the Mid-Atlantic Bight occurs later and results in the recruitment of young during August, at which time their primary prey is the bay anchovy.

LITERATURE. Roberts 1975; Freadman 1979; Safina and Burgers 1985; Juanes and Conover 1995

FISHERIES. All along the Atlantic coast bluefish is harvested by recreational and commercial fishermen. Bluefish is a favorite species for recreational fishermen because they bite greedily, put up a strong fight, and aggregate in schools and can be caught one after another.

Since 2000 the East Coast recreational fishery has harvested an average of 31.8 million pounds per year, although it was four times that level in the early 1980s. In New York the recreational harvest from 2000 through 2017 has averaged 7.1 million pounds. Since then, and during the rebuilding mode, New York recreational harvest has been 1.5 to 3.5 million pounds. The overall recreational harvest has been under a harvest limit since 2000.

As with many other coastal fishes that could be caught with simple inshore gear, the peak commercial harvest of bluefish occurred in the late 1800s and early 1900s. Peak commercial landings in New York occurred in 1904 when over 11 million pounds were landed. Since 2000 New York commercial landings have averaged 1.3 million pounds annually. Commercial landings of bluefish have been regulated by quota since 2000.

ALL TACKLE WORLD RECORD: 31 lb. 12 oz., Hatteras, North Carolina, 1972

NEW YORK RECORD: 25 lb. 0 oz., Montauk, 1998

MANAGEMENT. Bluefish is managed jointly by the MAFMC, NMFS, NYSDEC, and the ASMFC through the Bluefish FMP (MAFMC 1990). The Bluefish FMP was approved in 1989 and quotas were established in 2000. As of the writing of this book, bluefish is overfished but not currently experiencing overfishing (NMFS 2020). However, the FMP mandates that any overfished/overfishing determinations be immediately addressed, thus bluefish is in a rebuilding mode.

The management plan divides the total annual coastal quota between the recreational fishery (86%) and the commercial fishery (14%). Additionally, there can be annual quota transfers between the commercial and recreational sectors, in either direction, of up to 10% of the overall quota.

The recreational fishery is managed coast-wide without state-by-state quotas. However, each state must restrict the possession of bluefish by anglers to constrain harvest and rebuild the stock. As of the writing of this book, in New York anglers are limited to a total of three fish per day (including snappers) or five fish per angler on party/charter boats. There is no closed season.

For the commercial fishery the total commercial quota is divided among the states based on updated landings during the period 2009 through 2018. Based on this update, New York will be increasing its state quota from 10% to nearly 20% of the coast-wide commercial quota over a 7-year period starting in 2022. In New York there is a 9 in. total length minimum size in the commercial fishery. Additionally, the NYSDEC sets an adjustable trip limit to spread the catch out through the season and can periodically close the fishery if catch starts to approach quota. The majority of the coastal commercial bluefish harvest takes place in New York, New Jersey, and North Carolina. In New York most bluefish are landed by gillnet, pound net and trawl.

JACKS
Family Carangidae (*jureles*)

This is a highly diverse and widely distributed family with about 140 species. There are 28 species that occur to varying degrees along the U.S. coastline, although most occur in warm southern waters. Their bodies range from compressed to spindle shaped, many have scutes arranged along some portion of their lateral lines, and all have three anal spines with the first two detached from the other fin rays. The distribution of carangids in New York marine waters ranges from bays to nearshore and farther offshore. In the warmer months, juvenile African pompanos, moonfish, lookdown, banded rudderfish, and permits tend to be commonly found inshore.

African pompano

Alectis ciliaris (pompano de hebra)

Other common names: threadfin
Worldwide tropical and subtropical, juveniles from Massachusetts to southern
    Florida, adults to Gulf of Mexico, Brazil
Max. length = 4.9 ft., common = 2.9 ft.

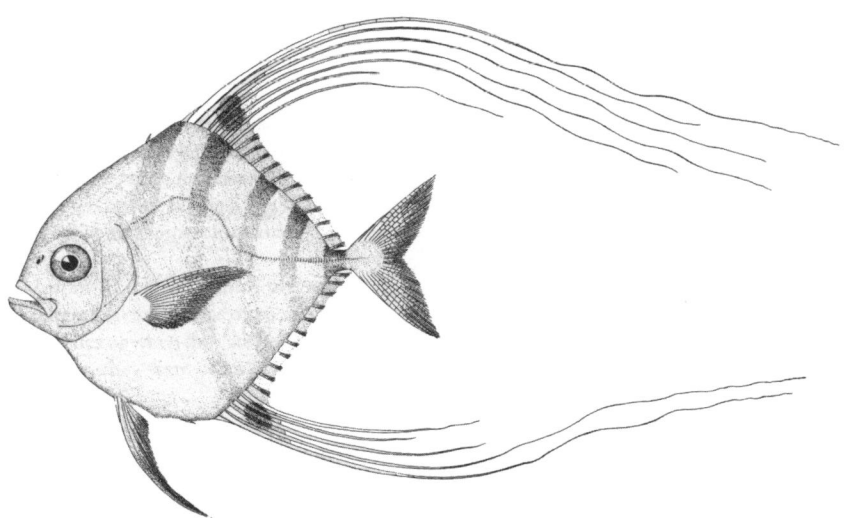

FIELD CHARACTERISTICS. Seen as juveniles locally with the first few soft rays of the dorsal and anal fins remarkably extended into very long filaments. These filaments are not seen in larger (over 16 in.) African pompanos.

Blue runner

Caranx chrysos (cojinuda negra)

Other common names: hardtail, yellow jack
Nova Scotia to Florida, Gulf of Mexico and Brazil, eastern Atlantic and
    Mediterranean
Max. length = 24 in., common = 14 in.

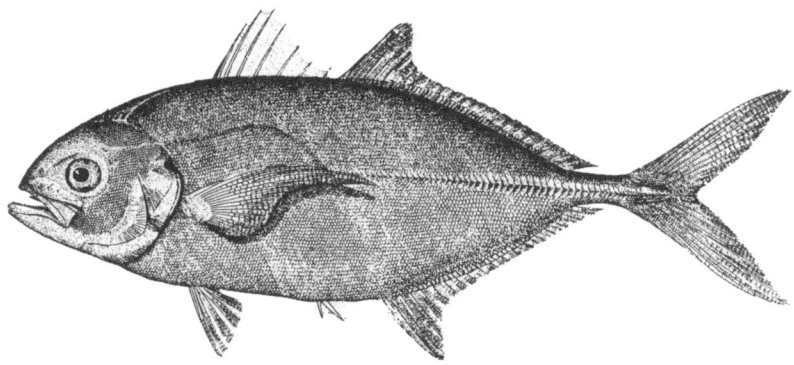

FIELD CHARACTERISTICS. This species has a typical jack shape and features, that is, a row of bony scutes along the back part of the lateral line, slender caudal peduncle and a forked tail, and two short, detached spines preceding the anal fin. The blue runner has a small black spot at the edge of the operculum and the tips of the caudal fin. Juveniles are barred on the side.

Crevalle jack

Caranx hippos (jurel comun)

Nova Scotia to Florida, Gulf of Mexico, Uruguay, eastern Atlantic and Mediterranean,
    rare north of Massachusetts
Max. length= 3.3 ft., common = 1.4 ft.

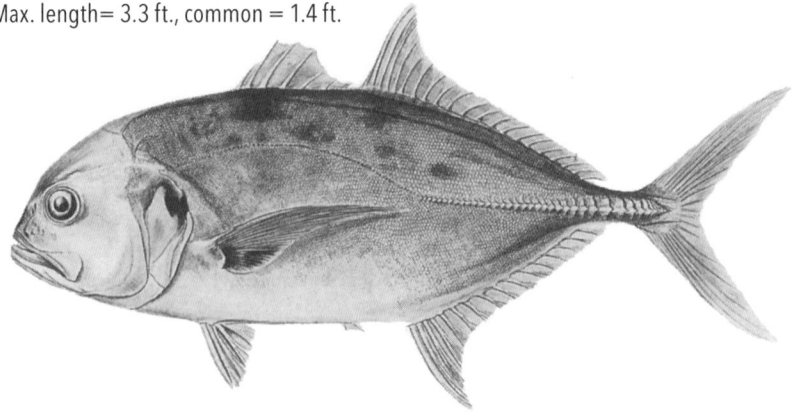

FIELD CHARACTERISTICS. This jack has a black spot on the lower portion of long pectoral fins and a black blotch on the operculum. Juveniles that will most likely be encountered locally have five to six bars on their sides.

Mackerel scad

Decapterus macarellus (macarela caballa)

Mostly tropical and subtropical from Bermuda to Brazil and rarely to Nova Scotia, also in the eastern Atlantic.
Max. length = 14 in., common to 8 in.

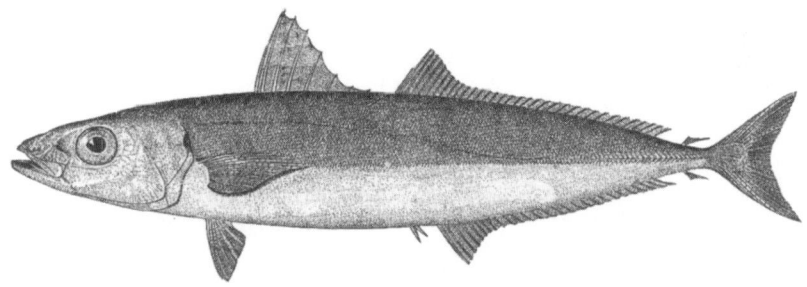

FIELD CHARACTERISTICS. There are small, detached caudal peduncle finlets between the second dorsal and the caudal fin as well as between the anal fin and the caudal fin. It also has a more elongate body than most jacks.

Round scad

Decapterus punctatus (macarela chuparaco)

Massachusetts to Florida, Gulf of Mexico, Brazil
Max. length = 9 in.

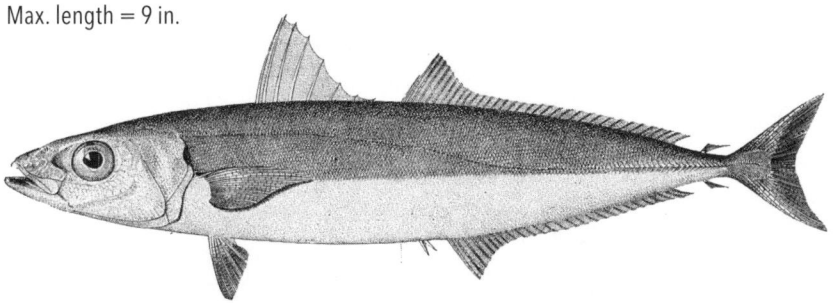

FIELD CHARACTERISTICS. Like the mackerel scad, the round scad also has a detached finlet placed near the caudal peduncle, but the round scad has a series of very small black dots on the curved portion of the lateral line.

Leatherjack

Oligoplites saurus (pina sietecueros)

Other common names: leatherjacket
Maine to Florida, Gulf of Mexico, Uruguay
Max. length = 11 in.

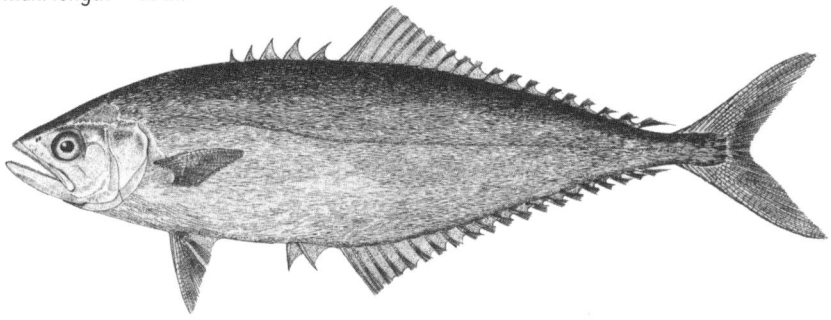

FIELD CHARACTERISTICS. There are no bony scutes along the lateral line, but it does have many detached finlets behind the dorsal and anal fins. The first five dorsal fin rays consist of five separate spines.

Bigeye scad

Selar crumenophthalmus (charrito ojon)

Worldwide tropical and temperate, Nova Scotia to Florida, Gulf of Mexico, Brazil
Max. length = 11 in.

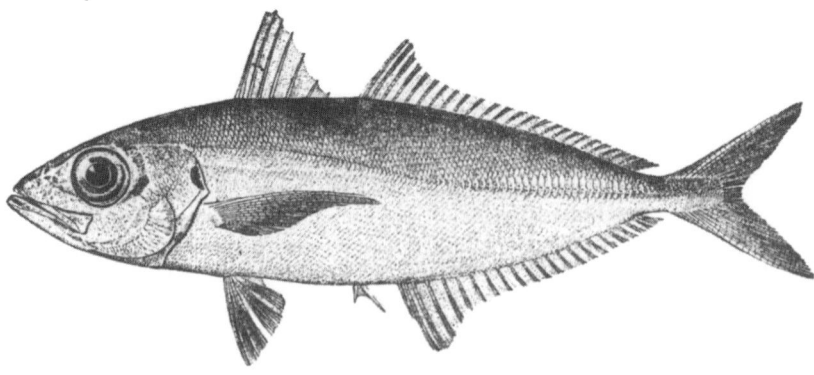

FIELD CHARACTERISTICS. As expected, the eyes are conspicuously large. There are no detached finlets behind the dorsal and anal fins.

Atlantic moonfish

Selene setapinnis (jorobado caballa)

Nova Scotia to Florida, Gulf of Mexico to Argentina, rare north of Cape Cod, common
    south of Chesapeake Bay
Max. length = 13 in.

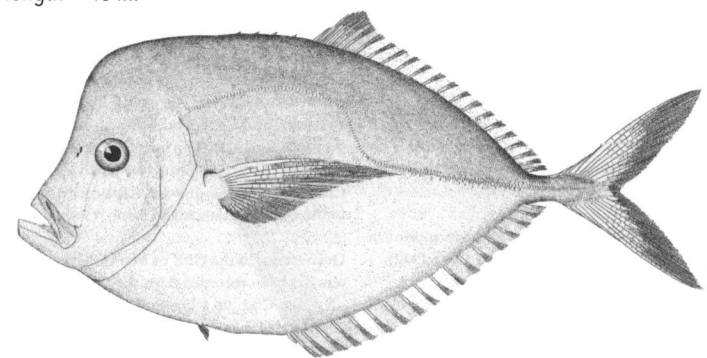

FIELD CHARACTERISTICS. The body is deep and highly compressed. The upper
anterior profile of the head is concave. In juveniles there is a moderately long
first dorsal fin filament that is reduced in adults and a dark midbody blotch
on its silvery sides.

Lookdown

Selene vomer
(jorobado penacho)

Cape Cod to Gulf of Mexico, Uruguay,
    rare north to Nova Scotia
Max. length = 18 in., common = 9 in.

FIELD CHARACTERISTICS. The silvery
body is deep and highly compressed.
The head is steeply sloping.
The second dorsal, anal, and
pelvic fins are long in juveniles.

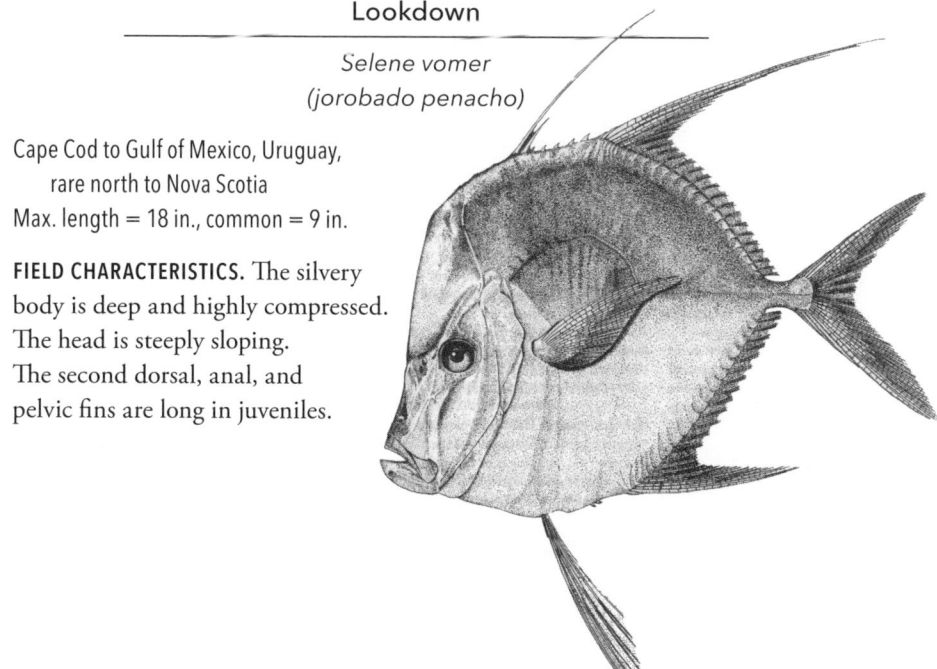

Greater amberjack

Seriola dumerili (medregal Coronado)

Nova Scotia to Florida, Gulf of Mexico, Brazil
Max. length = 5 ft.

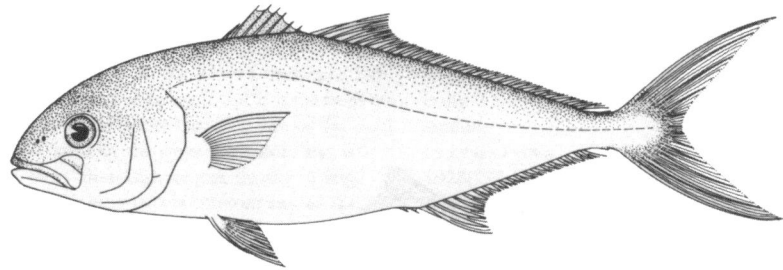

FIELD CHARACTERISTICS. The body is tapered and spindle-shaped. The pectoral fins are shorter than the head, the second dorsal fin is longer than the anal fin, and there are no finlets or scutes on the peduncle or elsewhere. Often there is a stripe running from the nose, through the eye, to the front of the first dorsal fin.

Banded rudderfish

Seriola zonata (medregal rayado)

Nova Scotia to Gulf of Mexico, and Brazil
Max. length = 2.2 ft., common = 1.5 ft.

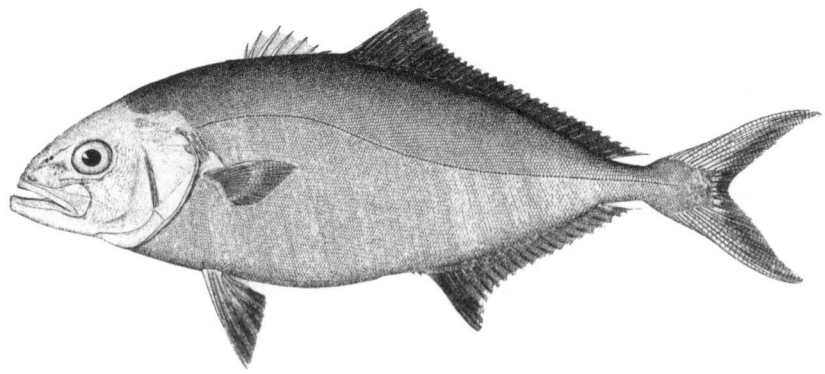

FIELD CHARACTERISTICS. Banded rudderfish are represented by conspicuously marked juveniles that have six broad dark bars on the trunk and dark bands through the eyes and on all the fins.

ECOLOGY/LIFE HISTORY. The juveniles are occasionally confused with rarely seen pilotfish, *Naucrates ductor*. When young, they are associated with floating debris and seaweed strands but may also be seen around jellyfish, sharks, larger fish, and occasional SCUBA divers. The dark bands slowly disappear in adulthood. As benthopelagic fish, the adults are found over hard substrates from inshore to greater depths feeding on crustaceans and fishes.

Florida pompano

Trachinotus carolinus (pompano amarillo)

Massachusetts to Florida, Gulf of Mexico, Brazil
Max. length = 2.0 ft.

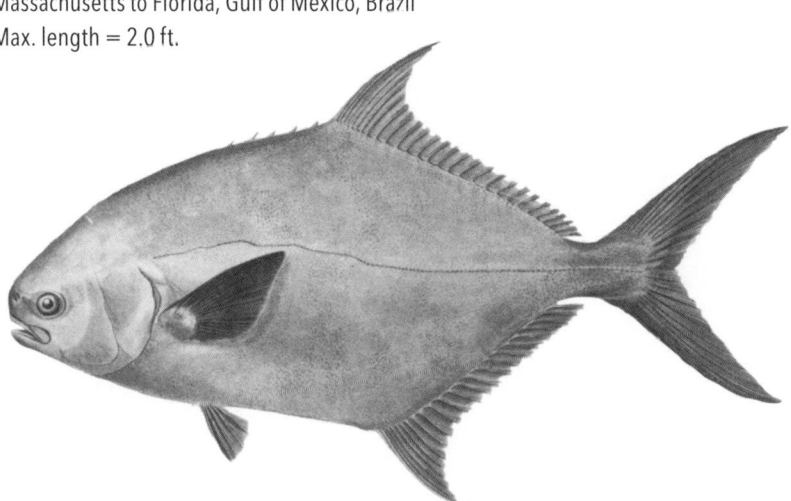

FIELD CHARACTERISTICS. Similar to the permit, but the Florida pompano has yellow caudal and anal fins. A short first spiny dorsal fin precedes a second dorsal fin. The second dorsal fin has 22–27 rays, more than that in the permit.

Permit

Trachinotus falcatus (pompano palometa)

Massachusetts to Gulf of Mexico, Brazil
Max. length = 2.6 ft.

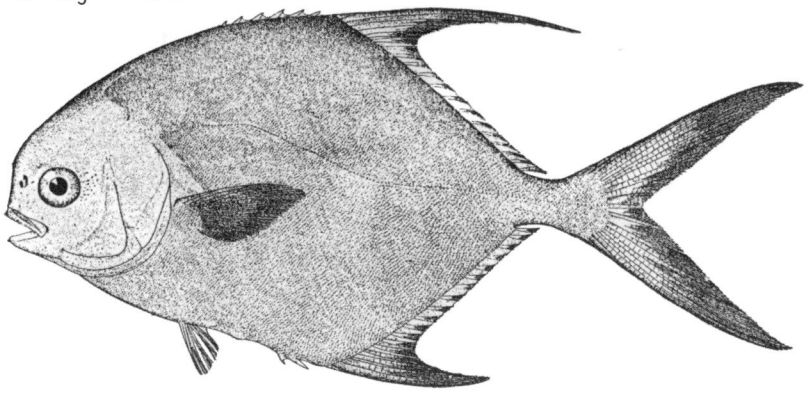

FIELD CHARACTERISTICS. Occur as juveniles with a rounded, compressed body, deeply forked tail, and long anterior parts of the dorsal and anal fins.

Rough scad

Trachurus lathami (charrito garreton)

Maine to Argentina, rare north of New York
Max. length = 13 in., common = 7 in.

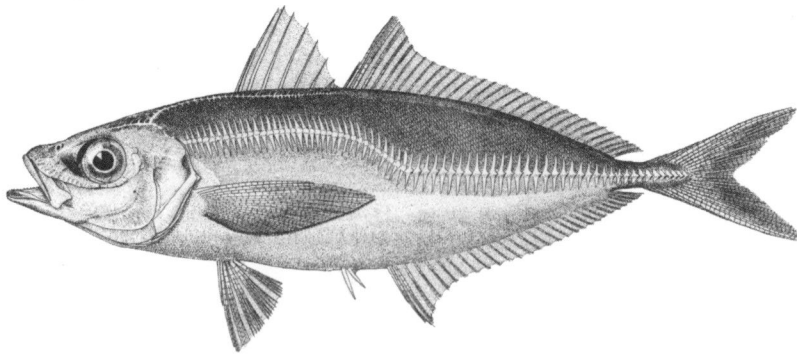

FIELD CHARACTERISTICS. There are bony plates along the entire lateral line from the margin of the gill cover to the caudal fin.

COBIAS
Family Rachycentridae (*cobia*)

There is only one species in this family, but it ranges throughout the warm temperate waters of the Atlantic and the Indo-Pacific region. Cobias resemble some species of remoras, especially when they are juveniles. As adults both fishes are elongate and have projected lower jaws and somewhat common geographic distributions. Unlike remoras, cobias do not have the first dorsal fin modified into a suction disc but do have a relatively unique set of short, isolated spines preceding a long second dorsal fin.

Cobia

Rachycentron canadum

Massachusetts to Argentina
Max. length = 6.5 ft.

FIELD CHARACTERISTICS. This fish most commonly occurs as a juvenile in local waters. It has a distinctive dark stripe running from its snout, through the eye, to the base of its caudal fin.

ECOLOGY/LIFE HISTORY. This species is mostly caught by anglers rather than by commercial fishermen. It is considered a good food fish.

DOLPHINFISHES
Coryphaenidae (*dorados*)

These fish are often just called "dolphins," but this name commonly leads to confusion, among the public, with the marine mammal. There are only two species in the family, and both are distributed worldwide in tropical and temperate waters.

Dolphinfish

Coryphaena hippurus (dorado)

Other common names: mahi mahi
Massachusetts to Brazil
Max. length = 6.5 ft., common = 3.2 ft.

FIELD CHARACTERISTICS. The profile of the head in mature males becomes vertical rather than rounded. The single long dorsal fin originates on the head. That fin and the long anal fin have no spines, only soft rays. The caudal fin is deeply forked. The erect dorsal fin is blue and all other fins are golden yellow. In life, the flanks of the compressed body are equally colorful.

FISHERIES. There is a minimal commercial fishery for dolphinfish in New York. Since 2010 landings have been less than 20,000 lb./yr. and are often less than 10,000 lb. There is a larger recreational fishery for dolphinfish, with highly variable New York landings ranging from a few hundred thousand pounds per year to less than 20,000 lb./yr.

MANAGEMENT. New York has no specific regulations for either the commercial or recreational dolphinfish fishery, and most fish are caught outside New York waters. In federal waters the fishery is managed by the South Atlantic Fishery Management Council and NMFS through the Dolphin and Wahoo FMP (SAFMC 2003). Management measures include a commercial quota, a recreational harvest limit, commercial trip limits, and recreational possession limits.

REMORAS
Family Echeneidae (*remoras*)

The family includes four genera and eight species and has a worldwide distribution in tropical and temperate seas. All members of this family spend time attached to large moving organisms or objects. This is made possible by the development and use of unique anatomical structures and adaptive behavior. The main structural feature that characterizes this family is the 10–28 slat-like, flexible plates that form a suction disc on the flattened, forward part of the head. This structure is capable of forming a powerful suction when applied to the surface of a host. Remarkably, this device develops from a set of transformed first dorsal fin spiny rays. In an effort to reduce drag while attached to a host, remora family members have streamlined bodies and, because their scales are small and cycloid, their body's surface is smooth.

In general, sharks seem to be the most common host, but certain remora family species may show other preferences or no preferences. The list of potential hosts includes sharks, rays, sea turtles, marine mammals, swordfish, marlin, and ships and floating objects. Instances of remoras attempting to attach to SCUBA divers have been reported.

There is some debate over the relationship between remoras and their hosts. On the one hand remoras may benefit the host by picking and removing parasitic copepods and necrotic tissue. It is assumed that if the host is large enough the drag produced by the attached remora is minor. The benefits to the remora are significant in terms of energy savings, protection from predators, and acquiring a possible food source. Symbiotic relationships are categorized as mutualism if both partners benefit. If one partner benefits and the other is indifferently affected, then the relationship is called commensalism. It would appear that the latter may be the best description in this case.

Remoras live in the surface waters of the ocean where their large hosts occur. Although most remoras are observed while attached, free swimming remoras are not uncommon. On several island nations, natives have developed the fishing technique of tying a line to the caudal peduncle of a captive remora, then releasing the fish, and hauling it back after the fish has attached itself to a host.

A HISTORY LESSON

This species is an original Linnean species. In 1758, Carolus Linnaeus published the 10th edition of his book *Systema Naturae*. It was he who established the science of taxonomy and developed the system of binomial nomenclature whereby he named every organism with a two-part label, the genus and species. The sharksucker is formally identified as *Echeneis naucrates* Linnaeus, 1758. This means that the genus has remained unchanged since the time it was identified in 1758, There are now over 28,000 species of fishes named to date, but only a few were identified in 1758 and still retain their initial name. Linnaeus (1707–1778) was primarily a botanist, but he is credited with naming the species listed in his book. However, most of the fish material and data he used was the work product of his colleague, Peter Artedi (1705–1735), who should be considered the first systematic ichthyologist.

Sharksucker

Echeneis naucrates (remora rayala)

Massachusetts to Uruguay, Mediterranean, eastern and southern Pacific
Max. length = 3 ft.

FIELD CHARACTERISTICS. The sharksucker has a dorsally placed suction device made of 18–28 plates and, as in other remora species, it has a projecting lower jaw and no swimbladder. The projecting lower jaw facilitates feeding on the surface of the host, and the absence of a swimbladder reduces the effort the sharksucker needs to keep attached to the host object. To detach from the host's surface, the sharksucker simply swims forward, and this releases the suction effect.

LITERATURE. Fulcher and Motta 2006

Color Plates

Alewife (above) • Banded rudderfish (below)

Black sea bass (above) • Tautog (below)

Northern puffer (blowfish) (above) • Northern puffer (blowfish) (below)

Atlantic menhaden (above) • Cunner (below)

Smooth dogfish (above)　•　Summer flounder (below)

Juvenile lionfish (above) • Lookdown (below)

Northern searobin (above) • Oyster toadfish (below)

Planehead filefish (above) • Spotfin butterflyfish (below)

Rock gunnel (above) • Scup (below)

Lined seahorse (above) • Short bigeye (below)

Rough silverside (above) • Snowy grouper (below)

Spot (above) • Striped bass (below)

Striped searobin (above) • Atlantic sturgeon (below)

Gray triggerfish (above) • Windowpane (below)

SNAPPERS
Family Lutjanidae (*pargos*)

The family has a worldwide tropical and subtropical distribution. There are 105 species in the family; 64 of those species are placed in the genus *Lutjanus*, and 10 of those occur at various locations along the East Coast of the United States. Some are collected, but only rarely, as juveniles in New York waters.

Gray snapper

Lutjanus griseus (pargo mulato)

Massachusetts to Florida and the Gulf of Mexico, Brazil
Max. length = 2.0 ft.

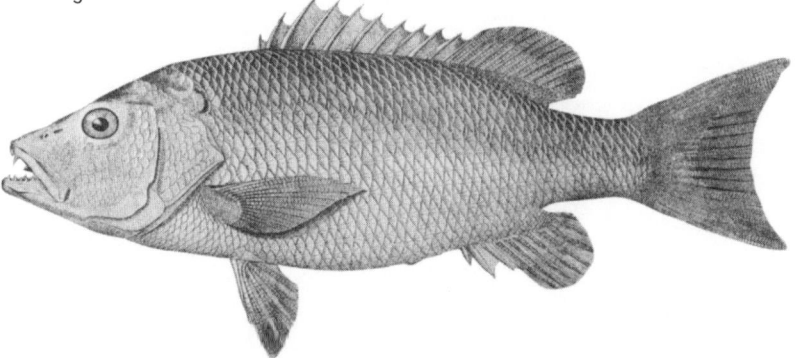

FIELD CHARACTERISTICS. It is likely that any New York gray snapper would be juvenile, so the fish would have a blue line below the eyes and a dark bar through the eyes. This and any other snapper specimen would have enlarged canine-like teeth and typical advanced teleostean features, such as a dorsal fin with a series of spiny rays followed by a set of soft rays, a pelvic fin placed below the pectoral fin, and scales that are ctenoid.

PORGIES
Family Sparidae (*plumas*)

Porgies are primarily marine fishes, although some species are occasionally found in brackish estuaries. Most species have a moderately deep, compressed body and a small, horizontal, and only slightly protractile mouth. The teeth at the front of the jaws are prominent and either conical or incisor-like; those at the sides of the jaws are molar-like. Most porgies are carnivorous, feeding principally on mollusks and crustaceans that they can crush with their molars. Some plant material is often included in the diet. Young fish form aggregations, but larger fishes are often less gregarious. Porgies are considered to be excellent food fishes, and many species are commercially important. The family, found in all tropical and temperate waters, includes 33 genera and about 115 species. Three of the latter are known to be in New York waters. The scup is the most common.

Sheepshead

Archosargus probatocephalus (sargo chops)

Maine to Florida, Gulf of Mexico and Brazil, most common from Cape Cod to Texas
Max. length = 3 ft.

FIELD CHARACTERISTICS. In contrast to the caudal fin of the more common scup, the outline of the caudal fin in the sheepshead is not lunate but is only slightly concave and the points of the fin are rounded rather than sharply pointed in the scup. The most conspicuous characteristic is the seven broad dark bars on the side of this fish.

Even though the species name is from the Greek for "sheepshead," it is unlikely that any similarity would be helpful in identifying the fish.

Pinfish

Lagodon rhomboides (xlavitia)

Cape Cod to the Gulf of Mexico
Max. length = 15 in.

FIELD CHARACTERISTICS. The common name of the pinfish refers to a small pin-like spine that is inclined forward and precedes the 12 spiny rays and 12 soft rays of the dorsal fin.

Pinfish have a distinctive black blotch near the origin of the lateral line, and the sides of the body usually have six dusky yellow bars and many blue and yellow longitudinal stripes. On closer inspection the anterior teeth are broad and not notched, unlike in those in the scup. These "rabbit-like" teeth likely gave rise to the name of the genus *Lagodon* ("hare tooth" in Greek).

The pinfish is an omnivore and uses those anterior teeth to pick at organisms that might be found on pilings and in marine meadows. Pinfish are becoming more common in the northern part of their range but have no significant commercial value in New York waters. Farther south, in the Gulf states, these fish are abundant and are caught as bait fish but are not considered a desirable food fish.

Scup

Stenotomus chrysops

Other common names: porgy
East Coast of North America, Nova Scotia to Cape Hatteras but rare
    north of Cape Cod
Max. length = 18 in., most adults < 14 in.

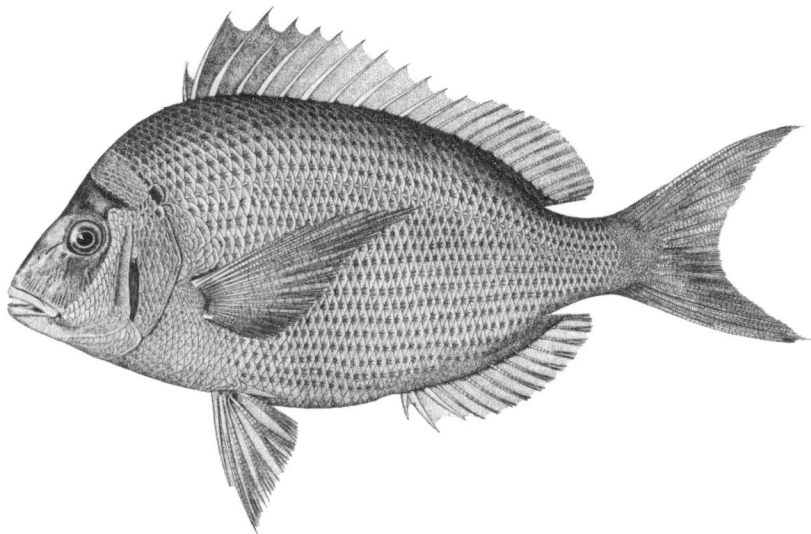

FIELD CHARACTERISTICS. There are some features, when taken as a group, that help to distinguish the scup from other fishes. This species has a single dorsal fin composed of spiny and soft rays. When the dorsal fin is erect, the spiny part is higher than the soft part. Further, the caudal fin is deeply lunate, that is, has a quarter-moon shape, the pectorals are long and pointed, and the dorsal profile of the head is slightly concave.

ECOLOGY/LIFE HISTORY. This fish is a continental shelf species that spends significant portions of its life in coastal waters. During the warm summer months, scup make a northern and inshore migration to sounds and tidal bays and usually congregate in schools. During the winter, an annual southern and offshore migration to deeper waters occurs. Local tagging studies show extensive inshore summering grounds located in New York waters and winter grounds offshore along the edge of the continental shelf from New York to North Carolina.

Spawning occurs in New York waters from May to August. Peak spawning occurs in June. Scup will feed on small crustaceans, worms, mollusks, and squid. The name *Stenotomus* has its origin from the Greek for "narrow cutting," referring to the narrow incisors used to grasp prey. Scup grow rapidly during the first 5 years of life and can live up to 15 years. Scup can reach a maximum weight of 5–6 lb., although most adults weigh less than 1 lb. Life history studies for scup show natural population fluctuations from periods of great abundance to periods of major scarcity. These unexplained natural population cycles and the highly migratory nature of scup complicate the development of management plans.

FISHERIES. All along the Atlantic Coast from North Carolina to Massachusetts scup is harvested by commercial and recreational fishermen. Scup continues to be one of the primary target species for commercial and recreational fisheries in New York.

Scup became increasingly important in New York commercial landings in the 1940s and 1950s, reaching a maximum of over 14 million pounds in 1958. The catch declined in the late 1960s and 1970s, then averaged between one and three million pounds until a quota system was implemented in 1997. Landings have been regulated by quota since then and have increased as the biomass of the resource has increased. New York commercial landings have averaged around three to four million pounds per year since 2010. The commercial fishery operates offshore in winter and inshore during late spring, summer, and early fall. The primary commercial fishing gear is the otter trawl, accounting for around 80% of the commercial catch. In the summer fishery, commercial fishermen also use pound nets and fish pots as well as hook and line.

Scup is a favorite species for recreational anglers because they bite greedily, tend to aggregate in schools, and can be caught one after another when anglers are on the right spot; also, they are a tasty pan fish. The scup recreational fishery occurs inshore. Recreational catches in New York fluctuated between a half million and nearly four million pounds between 1981 and 1996. As with the commercial catch, they have been limited by annual quotas (recreational harvest limit) since 1997. Since 2014 the New York recreational catch has averaged around 5.5 million pounds as the overall abundance of scup has been increasing.

ALL TACKLE WORLD RECORD: 4.5 lb., Nantucket Sound, Massachusetts, 1992

NEW YORK RECORD: 6.25 lb., 1978 (Note: One can speculate that the New York recordholder did not apply to the International Game Fish Association for world record status or the catch somehow did not qualify.)

MANAGEMENT. Scup is managed jointly by the MAFMC, NMFS, and by the ASMFC and the NYSDEC through the Summer Flounder, Scup, and Black Sea Bass FMP (MAFMC 1996). Scup is sustainably managed, and the resource was declared "rebuilt" in 2009. Additionally, as of the writing of this book, the resource is neither overfished nor is overfishing occurring and scup abundance is high (NMFS 2020). Scup were brought into the FMP in 1996, and quotas were established in 1997. The allocation between commercial and recreational sectors has recently been modified to 35% recreational and 65% commercial of the total annual quota. Recreational fishery management measures include a combination of minimum size limits, bag limits, and fishing seasons. Since 2004, the states of Massachusetts, Rhode Island, Connecticut, and New York have formed a northern region when setting their recreational regulations. This regional approach creates consistency between the states where fishermen from different states are often fishing alongside each other in the same waters.

The commercial quota is divided into three quota periods, winter I (January–April), summer (May–September), and winter II (October–December). A coast-wide quota regulates the two winter periods, while state-by-state quotas regulate the summer period. New York receives 15.8% of the summer period quota. Specific management measures for the commercial fishery include minimum size limits, minimum mesh requirements for trawls, and closure periods.

DRUMS AND CROAKERS
Family Sciaenidae (*corvinas y berrugatas*)

This is a large family with 270 species organized within 70 genera and distributed worldwide in waters ranging from fresh to brackish to marine. In North America there are 82 species although only 26 are found along the Atlantic coast from Canada to Mexico. The distinction they all share is the ability to produce sounds using swimbladder-associated muscles to drum against the bladder. Sound production and sonic muscle development in these fishes tends to be seasonal and limited to males. Although the species within this family vary in size and form, the lateral line in all of these extends on to the caudal fin.

Silver perch

Bairdiella chrysoura (roco amarillo)

Other common names: silver croaker
Cape Cod to Florida and Gulf of Mexico
Max. length = 11.8 in.

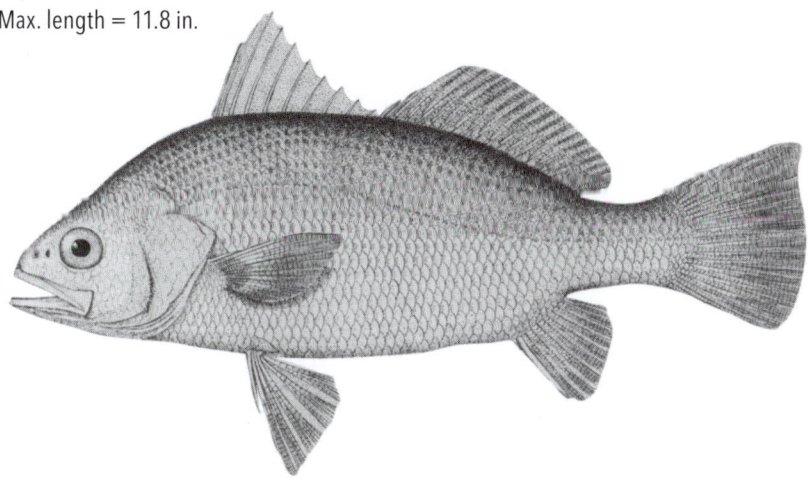

FIELD CHARACTERISTICS. The common name erroneously suggests that this species is related to perches, a large family of freshwater fishes. This confusion is not unusual when common names are used. The silver perch is primarily a coastal and estuarine fish although occasionally it might enter freshwater. The body color is silvery and the fins are yellowish (*chrysoura* = gold tail). There is no chin barbel, and the preopercular part of the gill cover is serrated.

Weakfish

Cynoscion regalis

Other common names: sea trout, gray trout, squeteague
Nova Scotia to Florida
Max. length = 3 ft.

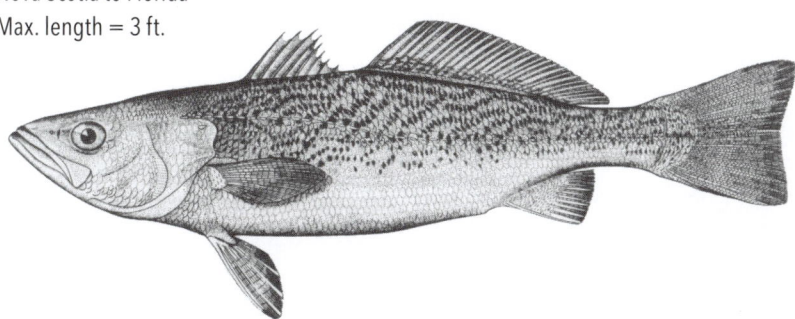

FIELD CHARACTERISTICS. The weakfish is a brilliantly colored fish with a greenish-blue dorsal surface, purple and lavender iridescent sides, and pelvic and anal fins that are pale yellow. The mouth is large and has two distinctive canines in the upper jaw. Although the jaws are not particularly weak, the common name of this fish is said to refer to the ease with which a hooked fish is lost when the jaw tears.

ECOLOGY/LIFE HISTORY. Adult weakfish overwinter offshore in waters south of Cape Hatteras and spawn in the estuaries along the East Coast from New England to Florida. Studies based on weakfish otoliths suggest that weakfish, like some anadromous fishes, have a strong homing tendency and return to their natal waters. As otoliths grow, each new layer of calcium carbonate captures the chemical signature, in the form of isotopes, of the surrounding water. Spawning site fidelity ranged from 60% to 81% of the spawners, and those that did stray did not miss by much. Although the homing mechanism is known in many migratory fishes, that mechanism is still to be determined in weakfish.

Weakfish are schooling fish and are relatively abundant during the summer. They feed on a variety of prey ranging from soft- to hard-bodied invertebrates and, in particular, small fishes such as anchovies, herring, menhaden, sand lance, and silversides.

LITERATURE. Connaughton and Taylor 1994; Thorrold et. al. 2001

FISHERIES. Weakfish has long been an important commercial and recreational species in New York and throughout the Mid-Atlantic. They have a tasty, mild flesh and put up a strong fight when caught recreationally and thus are fun to catch. However, reduced abundance has recently decreased their importance to our fisheries. Weakfish migrate into New York coastal waters in spring to spawn and then migrate south and offshore in fall. For those reasons, New York's commercial and recreational fisheries for weakfish take place spring through fall. The catch takes place primarily in state waters.

As with many other inshore easily caught species, New York commercial landings were highest in the late 1800s though the mid 1900s. Peak weakfish commercial landings occurred in 1908 at 11 million pounds, then were down to less than a half million pounds through the 1950s and 1960s. There was resurgence in landings to around 1.5 million pounds in the 1970s through the early 1980s before weakfish abundance precipitously declined coast-wide in the 1990s and into the twenty-first century. Commercial landings since 2010 generally been less than 50,000 lb. Weakfish is caught commercially with trawls, gillnets, pound nets, and seines.

Weakfish was a very popular recreational catch before abundance was reduced. Peconic Bay, Shinnecock Bay, and Great South Bay were productive recreational fishing areas where weakfish fed on grass shrimp in eelgrass meadows. New York recreational weakfish catches were as high as 1.7 million pounds in the early 1980s but have recently been 75,000 lb. or less.

ALL TACKLE WORLD RECORD: 19 lb. 12 oz., New York

NEW YORK RECORD: 19 lb. 12 oz.

MANAGEMENT. Weakfish abundance dropped off in the 1990s through the 2000s, and coast-wide abundance remains low. The stock is depleted (ASMFC 2019). However, the recent decline in abundance is not attributed to fishing mortality, which has remained relatively low. Natural mortality has increased considerably since the 1990s from factors such as predation, competition, and environmental stressors.

Weakfish is managed by the ASMFC and the NYSDEC through the Interstate FMP for Weakfish, which was implemented in 1985 (ASMFC 1985b). Continued abundance declines have resulted in stricter harvest controls. The commercial fishery is regulated by a minimum size and a very low bycatch trip limit (currently 100 lb.) and closed seasons. The recreational fishery is likewise severely restricted with a minimum size and currently a one fish possession limit. Similar commercial and recreational restrictions are in place coast-wide. However, even with these restrictions in place, abundance is not recovering.

Spot

Leiostomus xanthurus (croca)

Massachusetts to Florida and Gulf of Mexico
Max. length = 14 in., common = 10 in.

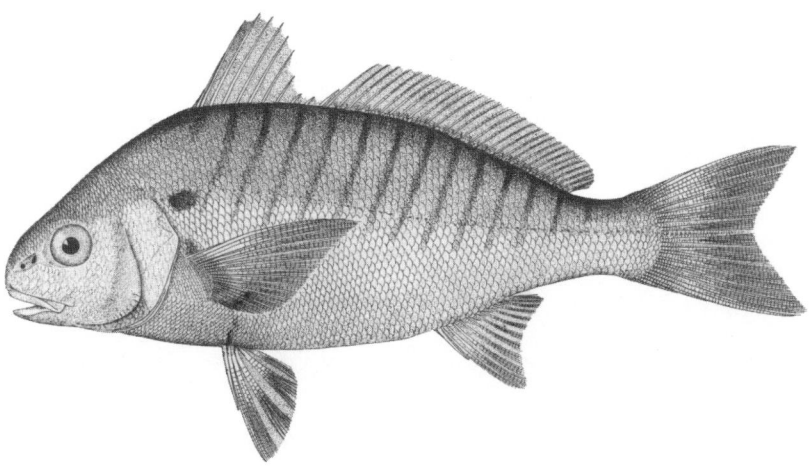

FIELD CHARACTERISTICS. In this species, the preopercular margin is smooth and there is a dark spot behind the upper angle of the gill opening. There are 12–15 oblique yellowish bands on the side and back and there is no chin barbel.

Northern kingfish

Menticirrhus saxatilis (berrugato)

Cape Cod to Florida, less commonly to Maine, most common Cape Cod to
    Chesapeake Bay
Max. length = 18 in., common =11 in.

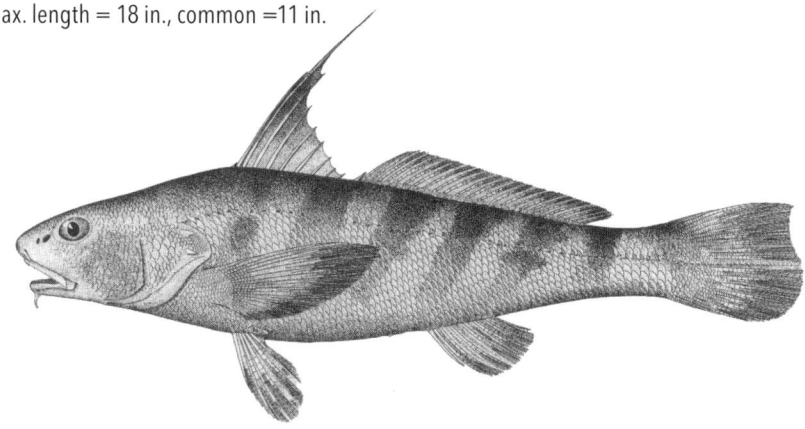

FIELD CHARACTERISTICS. There is a single chin barbel and the second and third dorsal fin spines are elongated. The blunt snout overhangs the mouth. The sides have many irregular broad dark bands and a V-shaped mark often appears on the fish's shoulder.

Atlantic croaker

Micropogonias undulatus (gurrubata)

Massachusetts to Florida to Gulf of Mexico
Max. length = 19.7 in.

FIELD CHARACTERISTICS. *Micropogonias* is Greek for "small beard" and refers to 3–5 pairs of minute barbels on the lower jaw. The middle of the body has short, irregular undulating brown streaks. There is a dark blotch at the base of the pectoral fin and the preopercle is serrated and spiny.

BUTTERFLYFISHES
Family Chaetodontidae (*peces mariposa*)

The family has 11 genera and about 122 species. Six species occur along the Atlantic coast of the United States, but only three occur in New York waters, as juveniles, seasonally.

This family is only one of many predominately warm-water fishes that occur as expatriate juveniles transported from the south by the Gulf Stream and end up in our bays during summer and early fall. As expected, these species do not survive in our cold winter waters nor do they emigrate south before hypothermia becomes a threat.

Butterflyfishes are the quintessential coral reef fishes and can be found throughout the world's tropical and subtropical regions. They are often the most conspicuous species found on shallow reefs where the fishes find secure habitats and a food supply. There are more colorful reef fishes, for example, angelfishes, parrotfishes, and wrasses, but butterflyfishes have very striking body patterns that distinguish each species.

These fishes are deep bodied, laterally compressed, and highly maneuverable within the complex reef habitat. They are often seen swimming in pairs. They have small, protractile mouths and small brush-like, sharp teeth. Small differences in each species dentition or snout length contribute to differences in their foraging behavior. Tubeworm tentacles, coral polyps, and sea anemones polyps are common diet items.

Foureye butterflyfish

Chaetodon capistratus (mariposa ocelada)

Massachusetts south to the Caribbean and to northern S. America
Max. length = 6 in.

FIELD CHARACTERISTICS. The "four eyes" referred to in the common name counts the pair of true eyes plus the distinct eye spot (ocellus) on each side of the body below the dorsal fin and just in front of the caudal peduncle. The sides of the body are white except for a series of lines running diagonally from the midline. Further, as is the case for the majority of butterflyfish species, there is a dark bar running down through the eye. Both the eye bar and the ocellus are meant to deceive predators as to the orientation of the butterflyfish.

Spotfin butterflyfish

Chaetodon ocellatus (mariposa perla amarilla)

Nova Scotia to Gulf of Mexico, Caribbean, and Brazil.
Max length = 8 in.

FIELD CHARACTERISTICS. This is the most common butterflyfish species found locally during the warm months. It has the broadest range of all the butterflyfishes. It has a distinct dark spot at the base of the soft-rayed dorsal fin. This spot may serve as an ocellus, but it is not as well developed as the ocellus in the foureye butterflyfish. The body of the spotfin butterflyfish is silvery and the fins are yellow.

Banded butterflyfish

Chaetodon striatus (mariposa rayada)

Massachusetts to Bermuda, the Caribbean to Brazil
Max. length = 6 in.

FIELD CHARACTERISTICS. The body is silvery with five dark bars, of varying widths, on the body and fins. Small juveniles may also have a spot on the dorsal fin.

LITERATURE. Neudecker 1989; McBride and Able 1998

WRASSES

Family Labridae (*doncellas y senioritas*)

The wrasses are the third largest family of marine fishes. (The goby and sea bass families rank higher.) The wrasse family has 68 genera and 453 species. Most of the wrasse species are found in the world's tropical waters, although there are about 22 species found in the eastern Atlantic and Mediterranean region. Only two species, the tautog and cunner, are found exclusively along the temperate western Atlantic coastline. The tautog and cunner both appear to have very thick "lips," and it is from that feature (*labrus* = lip) that the family acquired its name.

Tropical wrasses can be dramatically colorful and assume a wide variety of shapes and sizes as a sample of some tropical Pacific species (humpback wrasse, bird wrasse, harlequin tuskfish, clown coris, and redribbon wrasse) would suggest. Some tropical wrasse species are protogynous, changing from female to male with age and social influences. Commonly, that sex reversal is accompanied by a color change. Further, within this diverse family, there are species that bury in the sand or produce mucous cocoons at night and other species that act as cleaners and pick off ectoparasites and dead tissue from host fishes. These latter wrasses are mostly tropical species but some attempt has been made to employ the European goldsinny wrasse to remove sea lice from caged farmed Atlantic salmon.

Labrids have standard pectoral fins, but these are conspicuously used like oars to move the fish forward. This mode of swimming is called labriform locomotion and is very distinctive but not completely unique to the wrasse family. Parrotfishes, surgeonfishes, and butterflyfishes also employ this form of movement but it is unlikely that a snorkeler would confuse those fishes with wrasses.

Tautog

Tautoga onitis

Other common names: blackfish, white chin, tog
Nova Scotia to South Carolina
Max. length = 3 ft.

FIELD CHARACTERISTICS. Large adult males can be dark gray to black except for a light ventral area from the chin backward. Thus, the most frequently used common name of blackfish and to a lesser degree white chin reflect the adult male's coloration. Females tend to have a mottled brown pattern, and juveniles may be cryptically colored to blend into the sea grasses within which it tends to hide. Thus, as in other fishes, age, sex, and habitat influences the body color and patterns found.

The body of the tautog is often described as being stout. The genus, tautog, is derived from the Narragansett Indian term "taut-auog" which means "sheep's head" and refers to the fish's rounded forehead. Like many wrasses, the tautog has large conical teeth in the front of each jaw and crushing plates in the rear of the mouth. Those pharyngeal teeth consist of two sets of bones, the upper pharyngeal on the roof of the throat and the lower pharyngeal on the floor.

ECOLOGY/LIFE HISTORY. Labrids tend to be inactive at night, and that is true of the tautog although even during the day this nonschooling benthic fish can be observed not moving and leaning against objects within the rocky areas it generally tends to inhabit.

The tautog has a formidable dentition to feed easily on a variety of mollusks, and crustaceans. One of its preferred food interests is the blue mussel. It is clear that there is an advantage in having prey that is incapable of fleeing.

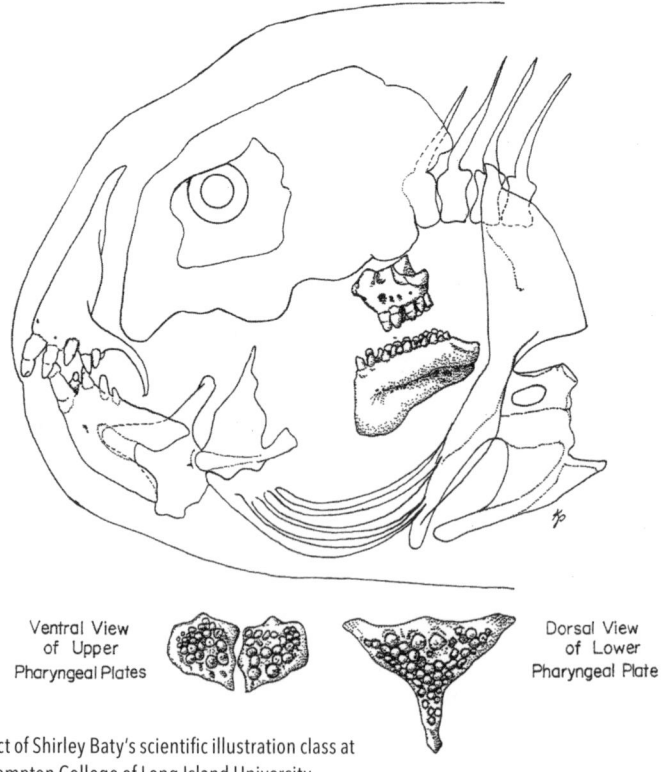

Ventral View of Upper Pharyngeal Plates

Dorsal View of Lower Pharyngeal Plate

Product of Shirley Baty's scientific illustration class at Southampton College of Long Island University.

As in most fishes, the geographic range within which a species has been observed historically is very broad, but in the tautog as in other species the range within which the fish is most likely to be collected is narrower. In this case, that latter range is from Cape Cod to Chesapeake Bay.

FISHERIES. Tautog is an important commercial and recreational species, particularly between Cape Cod and the Chesapeake Bay. Tautog have a tasty mild white flesh. They are relatively easy to keep alive and are often held and marketed live to promote absolute freshness and flavor and increased value. They put up a strong fight when caught recreationally and are fun to catch. Tautog tend to aggregate around structure and are caught around rocks, boulders, wrecks, jetties, and natural and artificial reefs. Tautog have strong site fidelity and inhabit the same location on a daily and annual basis. Tautog are caught commercially and recreationally inshore at shallow depths in the warmer months and then in deeper waters during the winter. Most catches occur in state waters.

Coast-wide, and in New York, recreational anglers catch around 90% of the total harvest, so the fishery is primarily a recreational one. Recreational harvest in New York peaked at four to five million pounds per year during the late 1980s before management regulations went into place. Since 2010 harvest has fluctuated widely between 500,000 lb. and three million pounds and is regulated by a recreational harvest limit. Tautog can be caught on a variety of baits but are particularly fond of live crabs. In early 2015 a New York resident caught a very large tautog in Maryland that broke the all-tackle world record. Most of the recreational catch takes place in New York, Connecticut, and Rhode Island.

Commercial landings of tautog were high in the late 1800s and early 1900s as with most other inshore easily caught species. Catches then dropped off to incidental during most of the twentieth century because of low market demand. Landings increased again to around 200,000 lb. during the 1980s and 1990s in response to increased demand from the live fish market as well as tautog being a bycatch of the lobster pot fishery, which was expanding rapidly during that time. Tautog is also caught commercially by rod and reel, pound net, fish pot, and trawl. After falling to a low of less than 50,000 lb. during the early 2000s, landings have again been 200,000 to 300,000 lb. and are regulated by quota with a trip limit.

ALL TACKLE WORLD RECORD: 28 lb. 12.8 oz., Maryland

NEW YORK RECORD: 22 lbs. 8.5 oz.

MANAGEMENT. Tautog is sustainably managed by the ASMFC and the NYSDEC through the Tautog FMP implemented in 1996 to reduce fishing mortality and rebuild the stock (ASMFC 1996). Tautog's slow growth rate and late maturity make it sensitive to overfishing. The FMP and the stock assessment have identified four regional stocks of tautog: Massachusetts–Rhode Island; Long Island Sound (Connecticut and north shore Long Island); New York–New Jersey Bight; and DelMarVa (Delaware, Maryland, Virginia, North Carolina). Owing to the influence of conditions that affect year-class strength as well as year-to-year variability of fishing mortality the overfished/overfishing status of the resource changes with updated stock assessments. However, the FMP mandates that any overfished/overfishing determinations be immediately addressed. As of the publication date of this book, tautog in Long Island Sound and the New York–New Jersey Bight are in a rebuilding mode (ASMFC 2021). The FMP has implemented actions to reduce fishing mortality and rebuild the stock, including quotas, trip and bag limits, seasons, and minimum size for both commercial and recreational fisheries. Additionally commercial fishermen must attach a metal tag to each fish on the opercula in order to verify that it was harvested by a licensed commercial fisherman. There is no federal management of tautog in waters outside of 3 mi.

Cunner

Tautogolabrus adspersus

Other common names: bergall, chogset
Labrador to Virginia
Max. length = 16 in., av. = 10 in.

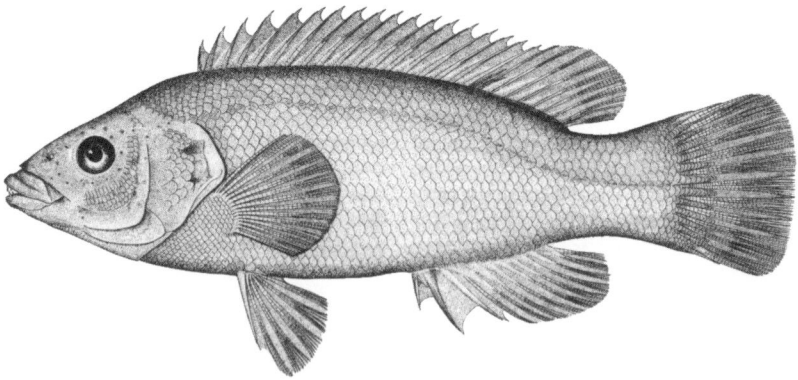

FIELD CHARACTERISTICS. The cunner is easily distinguished from the tautog. In the cunner, the gill covers are scaly, the snout is pointed, and it rarely exceeds 2 lb. In contrast, the tautog's gill covers are mostly naked, the snout is blunt, and the largest can weight over 20 lb.

It is not easy to describe the cunner's coloration in simple terms. Depending upon where it is found and its age, this species can have a uniform or a mottled color ranging from reddish brown to blue tinged with brown and juveniles living in eelgrass might be predominantly green.

ECOLOGY/LIFE HISTORY. Cunners are not schooling fish. Individuals generally associate themselves with structures such as pilings, jetties, rocky areas, and shellfish beds. It is on these surfaces where their major prey—mollusks, crustaceans, sea urchins, and worms—tend to concentrate. During the winter, cunners tend to move to deeper water and are less active.

Cunners have formidable canines located on the front of their extensible upper jaw (premaxillary) and on their strong low jaw (dentary). These teeth help to grasp their prey, but it is their pharyngeal teeth, in the throat, that serve to crush hard-bodied prey as it enters the esophagus. As food passes between them, the two components act as millstones to grind the ingested shelled material. Some shells are not completely broken down by this process, and because the cunner has no true stomach nor strong digestive acids, pieces of broken shells or whole shells go through the gut and are excreted unaffected.

LITERATURE. Olla et al. 1974; Liem and Sanderson 1986, Savaria and O'Connor 2013

EELPOUTS
Family Zoarcidae (*viruelas*)

The majority of the 230 species within this generally benthic family are in the North Pacific and North Atlantic, but 15 zoarcids are from Arctic Canada and 21 species are found in the Antarctic and subantarctic.

Ocean pout

Zoarces ameicanus

Labrador to Chesapeake Bay
Max. length = 3.8 ft., common = 1.8 ft.

FIELD CHARACTERISTICS. The body is very eel-like and smooth, mucus rich, and with small embedded scales. As in all family members, the long dorsal fin and anal fin are continuous with the caudal fin. The pectoral fins are very large but the pelvic fins are very small and placed under the throat in front of the pectoral fins. True eels have no pelvic fins and, in contrast to the eel pouts, the dorsal fin originates far behind the pectoral fins. The mouth is wide and ends behind the eyes. The sides of the body are marked with a chain-like, mottled pattern on a yellowish-brown background. Of the three species of this family that occur along our coasts, the ocean pout is most likely to be found nearer to shore. The other two species are the wolf eelpout (*Lycenchelys verrillii*) and the Atlantic soft pout (*Melanostigma atlanticum*).

GUNNELS
Family Pholidae (*espinosos*)

The family is distributed in the North Atlantic and Pacific with three genera and 15 species. These fishes are laterally compressed (ribbon-like) and elongate and have dorsal fins composed of more than 80 short stout spines. The anal fin is about half the length of the dorsal fin and has only two spines followed by soft rays. The pelvic fins may be absent or very small and placed just in front of the pectoral fin. In general, these fishes are not big and are often found in tide-pools or under rocks in shallow water.

Rock gunnel

Pholis gunnellus

Labrador to Delaware Bay and eastern North Atlantic Ocean
Max. length = 12 in., av. = 6 in.

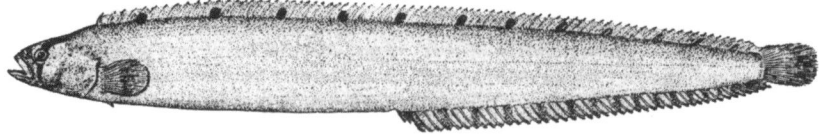

FIELD CHARACTERISTICS. This species is found amongst seaweeds and rocks and in the intertidal (littoral) zone.

The background color of the body is a light brown to yellowish, but there are darker brown bands and a distinctive dark brown stripe from the eye to the corner of the mouth.

The rock gunnel has minute pelvic fins and a row of about 10–14 round white-ringed black spots on the back just beneath the dorsal fin. The dorsal fin runs from just behind the gill covers to the base of the small caudal fin. This fish has scales but these are small, cycloid, and embedded. Further, the body has a thick layer of mucus. It could be suggested that the scales and the mucus provide a protective barrier against the hazards of moving within a rough, rocky environment. Because of their slippery surface and eel-like (anguilliform) swimming undulations, predators and collectors may find it challenging to capture these fishes.

SAND LANCES
Family Ammodytidae (*peones*)

The family can be found in the Arctic and along the coast of the Atlantic, Pacific, and Indian Oceans. There are eight genera and 23 species, and most members of this family possess an elongate body, a sharply pointed snout, a projecting lower jaw, minute cycloid scales lying between oblique folds of skin (lateral plicae), and a curious lateral skin fold along both sides of the abdomen. There are no spines in the single long dorsal and anal fins and no pelvic fins. The lateral line is placed just below the dorsal fin base, the jaws are essentially toothless, and there is no swimbladder. Further, the relatively high number of vertebrae (52–78) as compared with a more conventional fish, for example, the striped bass with 25 vertebrae, makes it possible for the sand lance's body to be very flexible. (The American eel has 102–112 vertebrae.) With the exception of the toothless jaws, the other structural characteristics found in sand lances are part of a complex of adaptations that produces a flexible eel-like body movement, is negatively buoyant, and can quickly and unhindered dive into a sandy bottom and maneuver the fish until completely buried.

American sand lance

Ammodytes americanus

Other common names: sand eel, sand lance
Labrador to North Carolina
Max. length = 6 in., av. = 4 in.

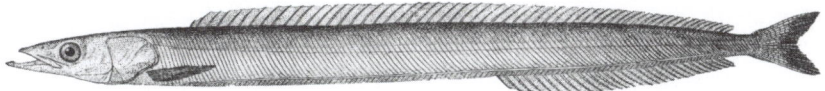

FIELD CHARACTERISTICS. The sand lance, like many fishes, is counter-shaded with a dark back and silvery sides and venter. The number of lateral plicae arranged along the sides ranges from 106 to 126. Although it might look like a member of the eel family, unlike an actual eel, the sand lance has a forked caudal fin.

ECOLOGY/LIFE HISTORY. These fish routinely bury themselves at sunset and remain buried until sunrise the following day. The name *Ammodytes* means sand diver. Being able to dive into the sand at any time provides protection against threats by predatory fishes (e.g., striped bass, bluefish, and cod) and sea birds (e.g., terns). These predators hunt during the daytime when schooling sand lances leave their sandy burrows to find and feed on plankton, worms, crustaceans, and small fishes. Although being buried is a reasonably effective antipredation method, some bottom-living fishes such as flatfishes and skates can detect and prey upon sand lances. Some observers have reported sand lances burying themselves between the high and low tide marks and remaining there even at low tide. Also, at times sand lances have been seen with their head protruding out of the sand.

LITERATURE. Meyer et al. 1979

STARGAZERS
Family Uranoscopidae (*miracielos*)

This family is found in the temperate regions of the Atlantic, Pacific, and Indian Oceans. It includes eight genera and 50 species.

As in most fishes, this family has a lateral line that permits them to detect vibrations in the water. The particular placement and development of the lateral line system is often related to a fish's habitat and habits. Stargazers burrow into soft substrate often with only the eyes, top of the head, and mouth exposed. In this case, the placement of the lateral line on the upper side of the body is adaptive because the unencumbered lateral line would still help the fish detect objects moving toward it.

A more unique anatomical feature involves the external nasal openings. In most fishes, the external nares are paired blind sacs that are used for olfaction. In this family, the external nares have tubular connections to the inside of the mouth. It is suggested that this unusual modification is also related to this fish's burrowing behavior. When buried, water to irrigate the gills for respiration cannot be pumped in through the mouth, so another entranceway is needed. This

method of using internal nostrils (choanae) is more commonly seen in the totally unrelated lungfishes and chimeras.

Another relatively unusual structural feature in this family is the presence
of two grooved venomous spines, each of which has a venom gland at its base.
These spines are just above the pectoral fin and behind the gill cover. The spine
itself is an extension of the supracleithrum, a bone that is part of the pectoral
girdle. As in most venomous fishes, the use of a painful toxin is for defense. The
northern stargazer is the most common venomous fish native to this region.
Other venomous fishes, a scorpionfish and a lionfish, are observed but are not
year-round residents and are relatively rare. If the venom doesn't deter predators, the electrogenic ability of this family will.

Northern stargazer

Astroscopus guttatus

Long Island to North Carolina
Max. length = 22 in., av.= 12 in.

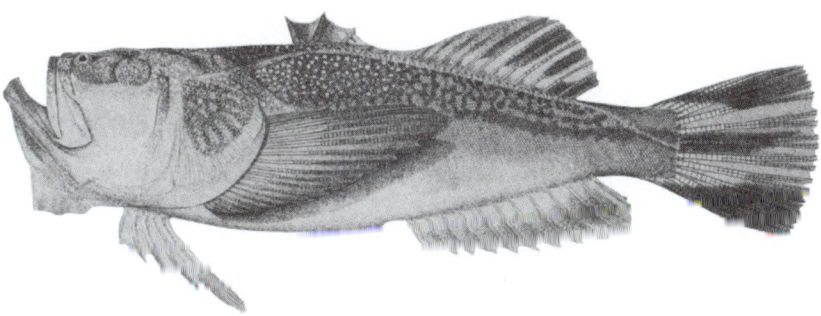

FIELD CHARACTERISTICS. As in the other members of this family, the northern
stargazer has a large head, with an oblique mouth and small eyes directed to
look upward. The name *Astroscopus* appropriately describes this appearance.
Behind the eyes, in members of the genus, and underneath the top of the
head, are two large electric organs derived from portions of muscles that
formerly controlled the movement of the eyes within the orbits. These electric
organs are capable of producing 50 V of electricity to be used to stun prey or
in defense against predators. This is the only marine bony fish that generates
enough voltage to stun prey. The only other marine fish that has a powerful
electric organ is the cartilaginous electric ray, *Torpedo*.

ECOLOGY/LIFE HISTORY. Stargazers are sedentary benthic fishes but are able
to feed on small fishes and crustaceans. The northern stargazer has a very
distinctive body pattern. The name *guttatus* means spotted and refers to the

many small spots on top of its head and body. Further, the large mouth, cryptic fringes around the mouth and nares, and the potential ability to stun prey give stargazers an advantage.

LITERATURE. Schwab 2004

COMBTOOTH BLENNIES
Family Blenniidae (*borrachos*)

Combtooth refers to the numerous, closely packed incisor-like teeth that are common in members of this family although some species have canines. The comb-like teeth are often used to graze on algae that cover the surfaces in the rocky inshore communities within which many live. Small marine organisms such as crustaceans, mollusks, and tunicates may also be in this family's diet. This is one of the more diverse fish families with 56 genera and about 360 species. Generally, they are benthic and live primarily in warm, shallow marine waters.

Feather blenny

Hypsoblennius hentz

Nova Scotia to the Gulf of Mexico
Max. length = 4 in.

FIELD CHARACTERISTICS. This species is typical of its family. It has a blunt head with small hair-like filaments (cirri) above the eyes. The reduced pelvic fins are placed in a jugular position in front of the large pectoral fins. It has a long single dorsal fin that contains about the same number of spiny rays as soft rays. Along the western Atlantic, its most common habitat is soft, muddy, oyster-shell-laden areas where the feather blenny can be found under rocks or shells. This is a secretive, mottled, slow-moving fish and would be difficult to discover unless one looks at the undersides of these rocks and shells.

GOBIES
Family Gobiidae (*gobios*)

This family is mostly marine and occurs throughout the world's warm temperate, subtropical, and tropical regions. The goby family has a greater number of genera (210) and species (1950) than any other fish family or vertebrate except the freshwater minnow family, Cyprinidae, which has 220 genera and 2420 species.

Gobies live in a wide range of habitats and often live in association with other animals, such as sponges, shrimp, and sea urchins. Some gobies are cleaners, picking ectoparasites from the skin, mouth, and gills of host fishes, and others are like mudskippers that are capable of being on land for short periods of time.

A distinguishing characteristic of this family is the united pelvic fins that form a modest but effective adhesive disc. Other less common fishes, clingfishes and snailfishes, have even more highly modified pelvic fins used for the purpose of adhering to benthic surfaces. That feature and their small size have contributed to gobies' ability to adapt to a variety of habitats that are not available to other fishes.

The members of this family are generally small, and, in fact, one of the smallest marine fishes is a goby, *Trimmatom nanus,* found on some reefs within the Indian Ocean. Mature females are only 8–10 mm in standard length. (Standard length is the straight-line distance between the tip of the snout and the base of the caudal fin but not including the caudal fin.) To date, the smallest marine fish is the stout infantfish, which belonged to a closely related family (Schindleriidae) native to the southern Pacific Ocean and has mature males with total lengths of only 7 mm. More recently that smallest species has been placed in the family Gobiidae.

Naked goby

Gobiosoma bosc (gobio desnudo)

Massachusetts to Florida and northern Gulf of Mexico
Max. length = 2.5 in.

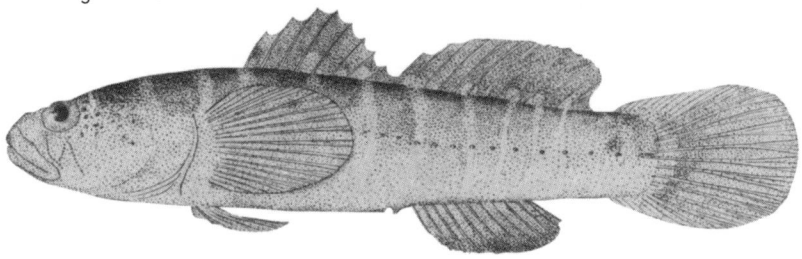

FIELD CHARACTERISTICS. The naked goby has no scales. This is not a unique distinction because all species of the genus *Gobiosoma* are "naked" with the exception of the seaboard goby, which has a pair of large ctenoid scales on each side of the caudal fin base. The naked goby's color is variable but generally greenish. There are 9–10 dark vertical bars on the sides separated by lighter narrower bars. (In describing body markings, vertical markings are called bars and horizontal markings are conventionally referred to as stripes.)

This species is more commonly found in shallow water, where it can live within grassy areas or empty shells. It feeds on small crustaceans and worms.

LITERATURE. D'Aguillo et al. 2014

Seaboard goby

Gobiosoma ginsburgi

Massachusetts to Georgia
Max. length = 3 in.

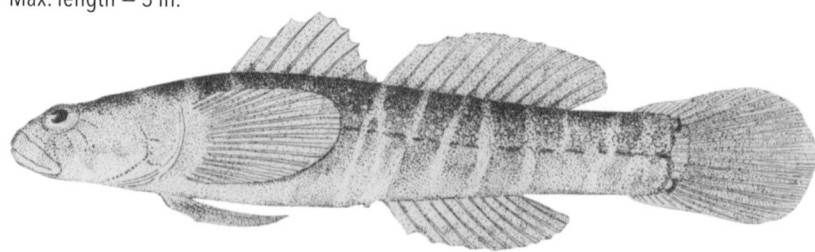

FIELD CHARACTERISTICS. Its body color is brown and its sides have bars but they are less distinct than in the naked goby. It lives in deeper water than the naked goby but also feeds on small crustaceans. This species is less common than the naked goby. Another difference between the two species is that the pelvic disc spans one-half the distance between the pelvic base and the vent in the naked goby whereas that distance is longer in the seaboard goby, spanning two-thirds the distance between the pelvic base and the vent.

SURGEONFISHES
Family Acanthuridae (*cirujanos*)

This is another large family of tropical marine fishes. It includes six genera and 80 species, but only 3 species are native to the reefs of the western Atlantic. For many of the tropical reef families, the vast majority of their species are found in the Indo-Pacific region. That region is considered to be the general area where the global coral reef fish fauna originated. The reasons for this assumption are associated with the large size of that area and its geological stability, thus providing many opportunities for speciation to occur.

Surgeonfishes are small, agile, and laterally compressed and are some of the most important members of the reef community because this family is one of the few marine families that is herbivorous. Other herbivores found in the Atlantic include damselfishes, parrotfishes, and chubs. It is notable that these families are common where light penetration and nutrient recycling promotes algal growth. In addition to harvesting the primary production of the reef, grazing by these herbivores prevents the algae from outcompeting the corals. One estimate of the percentage of marine species that are herbivores on tropical coral reefs is 30% or more in contrast to 10% or less in a temperate marine environment. Most of our temperate plant eaters feed on phytoplankton rather than macroalgae.

Surgeonfishes can forage as individuals but often will school with either only their species or as a mixed school with one or two other species. Surgeonfishes have small mouths but have many close-set and denticulate teeth designed to browse or graze on filamentous algae. A browser, like the blue tang, will remove parts of the plant whereas grazers like the ocean surgeon and the doctorfish bite the plant off at its base. These latter surgeonfishes use their gizzards to break up and then digest the algal material.

The most distinctive anatomical feature of this family, and from which it derives its name, is the lancet-like retractable spine on each side of the caudal peduncle. This spine folds into a horizontal groove. In some species, the spine is made more conspicuous by being a contrasting or darker color. The function of the spine is for defense and is extended whenever a threatening object approaches from that side.

Doctorfish

Acanthurus chirurgus (cirujano rayado)

Massachusetts to Florida, Gulf of Mexico, Caribbean to Brazil
Max. length = 14 in.

FIELD CHARACTERISTICS. This species has a highly variable body color but always has body bars although they might be faint. The tail is slightly indented and often has a white bar at its base.

Blue tang

Acanthurus coeruleus (cirujano azul)

Long Island to Florida, Gulf of Mexico, Caribbean to Brazil
Max. length = 15 in.

FIELD CHARACTERISTICS. As a juvenile (1.5–2.5 in.) the blue tang is bright yellow; as a slightly larger specimen (3–4 in.) its body color is blue but it has a bright yellow tail. As an adult (5 in. or larger), the body is blue with a yellow spine on the caudal peduncle. Many reef fishes assume different colors and patterns as they develop from younger to older individuals.

Ocean surgeon

Acanthurus tractus (cirujano pardo)

Massachusetts to Florida, Gulf of Mexico, Caribbean to Brazil
Max. length = 15 in.

FIELD CHARACTERISTICS. This species has a uniform pale blue or dark brown body color but with no bar pattern. Its caudal fin is crescent shaped.

BARRACUDAS
Family Sphyraenidae (*barracudas*)

This family has one genus with 21 species. It is widely distributed within the warm waters of the Atlantic, Pacific, and Indian Ocean regions.

Elongate bodies, large mouths, and strong teeth characterize this family of carnivores. As a general rule carnivorous fishes have poorly developed gill rakers, and this is true for this family. Gill rakers are of particular value only to planktivores who use them to filter particulate food matter, which is then directed to the esophagus and digestive tract.

In most fishes with two dorsal fins, the first one consists of spiny rays and the second one is usually soft rayed. This is the case for barracudas. However, another general feature is that fishes with two dorsal fins usually have those fins very close together if not touching. This is not true of barracudas and a small number of other fishes, for example, the mullets.

The most notable member of the family is the great barracuda, *Sphyraena barracuda*, but the family is represented locally by only its smaller relative, the northern sennet.

Northern sennet

Sphyraena borealis

Other common names: northern barracuda
Nova Scotia to Florida, Gulf of Mexico, Panama
Max. length = 18 in., av. = 12 in.

FIELD CHARACTERISTICS. A sennet is a trumpet or cornet used to signal the exit or entrance of sixteenth-century British stage actors. The common name likely refers to the shape of the fish. They are found on a variety of bottoms offshore and as juveniles are commonly found inshore.

Even though this species is a relatively small member of its family, the northern sennet is still an effective predator on smaller fishes, squid, and shrimp. This fish relies upon a burst of speed to overtake its prey. The northern sennet shares with other barracudas, needlefishes, and pike body features such as a flexible body and a large caudal fin that promote fast starts. Thus, it is able to lie in wait until the prey is within capture distance.

MACKERELS AND TUNAS
Family Scombridae (*macarelas*)

There are 16 species within this family that occur within the boundaries of the East Coast of the United States. In total, 51 species can be found worldwide in tropical and temperate seas. The majority, 11 of 16, exceed lengths of over 3 ft.

In general, fishes in this family have strong bodies, are active, swim fast, school, and some migrate over great distances. As in many fast-swimming fishes the body is designed to reduce drag. In addition to streamlining, the scombrids have a suite of adaptions that promote swimming efficiency. Their dorsal fins as well as pectoral fins in some scombrids insert into grooves when swimming quickly. The skin is smooth but has very small cycloid scales. Further, several short nonretractable finlets (5–12) occur behind the second dorsal and the anal fins. The function of these finlets is still uncertain. Some authorities hypothesize that the finlets eliminate vortices in water shed from the dorsal and anal fins and thus provide a less turbulent environment for the oscillation of the tail. Experiments that compared a tuna with and without finlets concluded that finlets were unlikely to function as significant drag reduction and thrust enhancing devices in routine steady swimming. However, it is possible that finlets may improve performance at high sustained speeds in rapid acceleration and turns. All the tuna species have narrow caudal peduncles and two small keels at the base of a strongly lunate caudal fin. The caudal fin constitutes a powerful propeller that can drive the body forward. In tunas, the bulk of body musculature is used to generate rapid lateral movements of the caudal fin rather than undulations of the body. About 90% of the thrust produced is contributed by the caudal fin.

Scombrids are generally epipelagic fishes; that is, they tend to live in the "photic zone" between the surface and 650 ft. However, there are instances in which these fishes have been recorded to dive to much deeper depths.

Little tunny

Euthynnus alletteratus (bacoreta)

Other common names: false albacore, albie
Nova Scotia to Florida, Gulf of Mexico to Brazil, Mediterranean, not common north
of Cape Cod
Max. length = 3.5 ft.

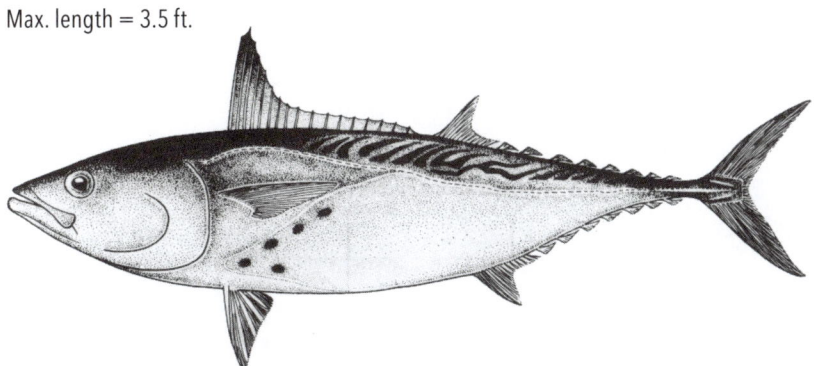

FIELD CHARACTERISTICS. Several black spots present between the pectoral and pelvic fin bases. The back is a dark green with wavy lines.

ECOLOGY/LIFE HISTORY. Like many tunas, the little tunny has no swimbladder and thus must keep moving to avoid sinking. When it needs to it can swim at very high speeds (40 mph). The ability to swim fast and for long periods of time without fatigue is, in part, a function of how much red muscle is available to drive the propulsive movements of the caudal fin. Little tunny tends to chase and feed on schools of small fishes. As in most carnivorous fishes, the little tunny has a short intestinal tract. A longer alimentary tract would be required if this fish needed to digest plant material.

Unlike most tunas, this species often occurs in relatively shallow water nearshore so it is accessible to surf fishermen. However, in general, the fish when caught is not eaten. The thick dark red muscle arranged along the length of the body around the spine is unpalatable unless the fish is quickly bled and iced. Otherwise, this catch makes excellent bait.

LITERATURE. Dickson 1995

FISHERIES AND MANAGEMENT. See the sections below on Tuna Fisheries and Management of Tunas, Billfishes, and Swordfish for details on fisheries and management of this species.

Skipjack tuna

Katsuwonas pelamis (barrilete listado)

Other common names: striped bonito; ocean bonito
Tropical and warm waters worldwide, both sides of the North and South Atlantic, in Mediterranean
Max. length = 3.8 ft., common = 1.4–2.6 ft.

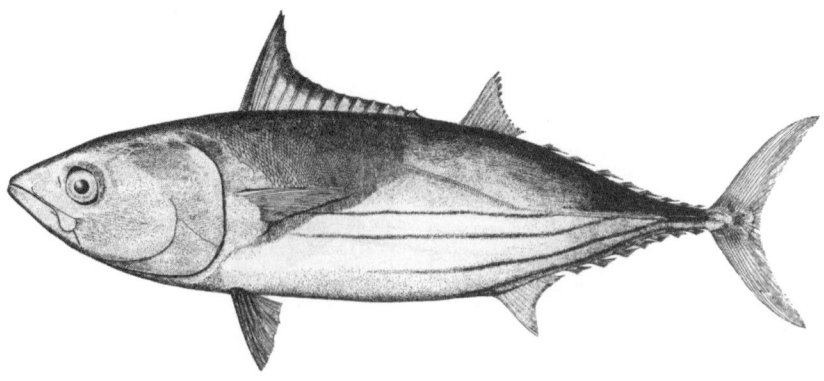

FIELD CHARACTERISTICS. The dorsal area is bluish black and the upper part may have some iridescent bands. There are four to six prominent dark and longitudinal stripes on the lower part of the body.

ECOLOGY/LIFE HISTORY. In contrast to offshore tunas, for example, yellowfin and bigeye, the skipjack is found more often in coastal waters.

It is an epipelagic species, that is, it lives in the "photic zone" from the surface to 650 ft. although the skipjack is able to go deeper. It has no swimbladder and can change depth without the need to alter the gas pressure in a swimbladder. Skipjacks swim in large schools and feed on such forage fishes as herring, anchovies, and sardines.

FISHERIES AND MANAGEMENT. See the sections below on Tuna Fisheries and Management of Tunas, Billfishes, and Swordfish for details on fisheries and management of this species.

Atlantic bonito

Sarda sarda (bonito del Atlántico)

Other common names: common bonito; inshore bonito
Nova Scotia to Argentina, Scandinavia to South Africa, Mediterranean
Max length = 3.5 ft., common = 2 ft.

FIELD CHARACTERISTICS. Compared with other "tunas," the Atlantic bonito is small. There are 5 to 10 longitudinal dark oblique stripes on the upper part of the body. The maxillary bone (upper jaw) is long and extends behind the posterior margin of the eye orbit.

ECOLOGY/LIFE HISTORY. Food includes other schooling marine organisms such as herring-like fishes, silversides, sand lances, and squid.

FISHERIES AND MANAGEMENT. See the sections below on Tuna Fisheries and Management of Tunas, Billfishes, and Swordfish for details on fisheries and management of this species.

Atlantic chub mackerel

Scomber colias

Nova Scotia to Venezuela
Max. length = 1.4 ft., common length = 1 ft.

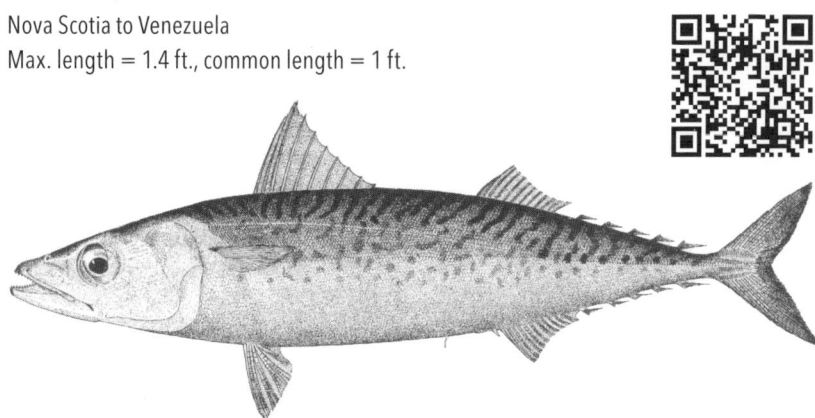

FIELD CHARACTERISTICS. As in the closely related Atlantic mackerel, the body is distinctly counter-shaded, having a blue back with many dark wavy lines fading into a lower side distinguished by dark blotches.

ECOLOGY/LIFE HISTORY. Curiously, unlike the closely related Atlantic mackerel, the Atlantic chub mackerel has a functional swimbladder. The latter species has been reported to feed at night but spend the day near the bottom, thus theoretically avoiding a need to regulate swimbladder gases at that time. Often, whether a fish has a swimbladder or does not is clearly related to whether it is exclusively benthic or whether it routinely changes depths as it feeds.

Atlantic mackerel

Scomber scombrus

Other common names: tinker
Labrador to North Carolina and in the eastern North Atlantic to the Mediterranean
Max. length = 2 ft., common length = 1 ft.

FIELD CHARACTERISTICS. The body is slenderer than that of a typical tuna, and the tail in the mackerel tends to be more forked than lunate. The two dorsal fins are widely separated by a distance that is greater than the length of the base of the first dorsal fin. Another distinction is the presence of transparent adipose eyelids immediately in front of and behind the eye. These cover the eye except for a vertical slit over the pupil and may simply serve a protective function or also have useful optical properties.

The dorsal surface is iridescent dark blue. The upper sides have many dark oblique lines, and there is a broken dark stripe along the lateral line and below that the body is silvery.

ECOLOGY/LIFE HISTORY. Mackerels swim in large, dense schools. They may also school with herring, alewives, and shad. One mid-nineteenth century captain reported a mackerel school that was 0.5 mi. wide and 20 mi. long. Mackerels generally swim near the surface and feed on large zooplankton, squid, and small fishes. However, they often dive to depths that exceed 500 ft. Unlike many tunas (e.g., yellowfin and bluefin) the Atlantic mackerel does not have a swimbladder that would otherwise tend to limit the extent to which it can change depth quickly. Bluefins, however, have been recorded to make very deep dives.

During the winter, mackerel are found offshore on the continental shelf. In the spring, they migrate coastward to spawn. Mackerel mature very quickly, an average of 1.9 years for both sexes. For comparison, bluefin tuna females mature at 4–5 years.

Compared with bluefin tuna, the Atlantic mackerel is considered a healthier seafood choice. The mackerel has more omega-3-fatty acids (2.6 vs. 1.6 gm of fatty acid/100 gm of tissue). Further, mackerel contains relatively low levels of mercury whereas the bluefin can have much higher levels.

FISHERIES. Mackerel was more important to New York's commercial fishery during the early and mid-1900s than it has been in the past 50 years. This change has more to do with market demand than resource abundance. Mackerel has fallen out of favor with American consumers (mackerel have a very strong flavor), so domestic demand is low. Foreign demand is high, but exports from the United States vary considerably as Atlantic mackerel are caught on both sides of the Atlantic and European fishermen can supply much of the European market. Peak New York commercial landings occurred in the late 1940s at 3.7 million pounds and have generally averaged less than a half million pounds during the past couple of decades. Mackerel have always been a relatively cheap fish with a low return to fishermen. During the 1900s mackerel was landed in New York by pound nets and trawls. In the twenty-first century most of the landings are by trawl, as mackerel no longer migrate

EFFICIENT BREATHING

In scombrids there is a progressive increase in reliance upon ram-gill ventilation. Mackerels utilize this at only fast cruising speeds whereas *Thunnus* species are obligate ram-gill ventilators. The conventional form of pumping water over the gills (buccal gill ventilation) requires synchronous expansion and contraction of the buccal (mouth) and opercular (gill) cavities. However, tunas have nearly lost the ability to irrigate the gills by pumping, in part because of the firm attachment in tunas of the major bones of upper jaw, the premaxilllary, and maxillary bones. Buccal gill ventilation requires 15% of the total energy expended by a fish, thus ram-gill ventilation serves as a significant source of energy conservation.

LITERATURE. Roberts 1978

to inshore New York waters as they have in the past. Changing environmental conditions have altered the distribution of Atlantic mackerel, shifting the stock northeastward.

There has likewise been only a modest recreational mackerel fishery in New York. Mackerel can be fun to catch and can be caught several at a time on multiple-hook mackerel jigs, but they are not a popular gamefish.

ALL TACKLE WORLD RECORD: 3lb. 8 oz., Spain

NEW YORK RECORD: None

MANAGEMENT. Mackerel is managed by the MAFMC and NMFS through the Atlantic Mackerel, Squid, and Butterfish FMP (MAFMC 1978). However, owing to the influence of conditions that affect year-class strength as well as year-to-year variability of fishing mortality the overfished/overfishing status of the resource changes with updated stock assessments. However, the FMP mandates that any overfished/overfishing determinations be immediately addressed. As of the publication date of this book, mackerel is considered overfished with overfishing taking place and is in a rebuilding mode (NMFS 2022b). The federal waters (outside 3 mi.) commercial management measures incorporate an overall quota, trip limits for some vessels, and limited permits. In 2022 a bag limit was enacted for federal waters recreational management measures. New York does not have any commercial or recreational management measures on mackerel within New York waters (within 3 mi.).

Spanish mackerel

Scomberomorus maculatus (sierra comun)

Nova Scotia to Florida, Gulf of Mexico, not common north of Cape Cod
Max. length = 2.5 ft.

FIELD CHARACTERISTICS. The sides of the body have bronze spots but do not have longitudinal dark streaks as on the cero, a less-common, closely related species.

FISHERIES. The commercial and recreational fisheries in New York for Spanish mackerel are very small and inconsequential. Both fisheries generally land less than 5000 lb./yr.

Albacore

Thunnus alalunga (albacore)

Other common names: longfin white tuna
Worldwide, Nova Scotia to Florida, Argentina, not common in the Gulf of Mexico
    and not common north of Cape Cod
Max. length = 4.4 ft.

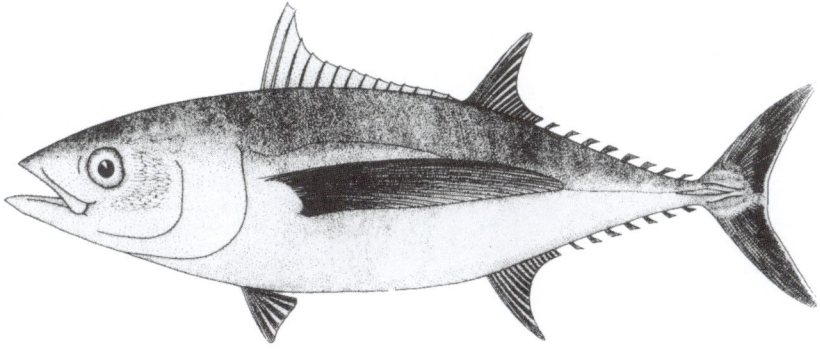

FIELD CHARACTERISTICS. The dorsal surface is dark and the pectoral fin is long, reaching beyond the origin of the second dorsal fin base.

ECOLOGY/LIFE HISTORY. Albacore tend to school with other tunas such as bluefin, yellowfin, and skipjack tunas. Large floating objects attract tuna schools. These floating structures can be natural, such as *Sargassum* seaweed masses, or fish aggregation devices. Both attract small prey that use the objects as shelter and that aggregation is exploited by predatory tunas. Aggregation devices are often used in the Pacific to catch pelagic tunas.

Albacore's most common prey are squid. That diet may or may not contribute to albacore's white meat, but the "white meat tuna" found in canned tuna is albacore.

FISHERIES AND MANAGEMENT. See the sections below on Tuna Fisheries and Management of Tunas, Billfishes, and Swordfish for details on fisheries and management of this species.

Yellowfin tuna

Thunnus albacares (atun aleta amarilla)

Other common names: Allison's tuna
Worldwide tropical and subtropical, not in Mediterranean
Max. length = 7.5 ft.

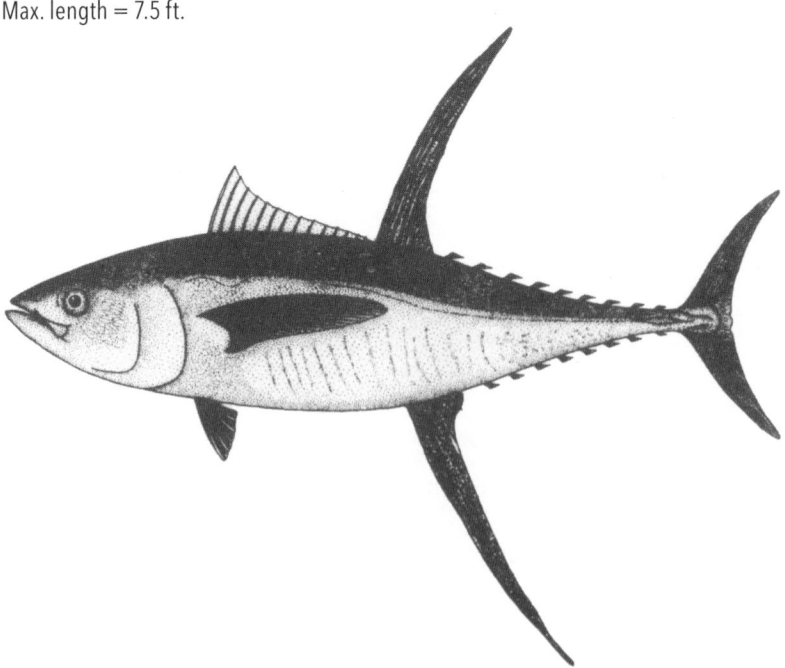

WARM-BODIED FISHES

In tunas and some sharks, cool blood leaving the gills is diverted to large peripheral blood vessels (cutaneous vessels) that run along the sides of the body and are the sources of the vascular system where counter-current heat exchange occurs.

This vascular system is a highly efficient heat exchanger and thermal barrier. In the skipjack, 95% of the heat is exchanged and thus conserved. In yellowfin tunas, muscle contraction power is doubled with every 50 °F (10 °C) increase in muscle temperature and this heat exchange contributes to higher swimming velocities and allows these fishes to swim farther and to outswim prey. This ability to use internally generated heat to maintain a warm temperature in muscles, gut, brain, and eyes is called heterothermy, and, in addition to tunas, the ability is exhibited by thresher and mackerel sharks, swordfish, and marlins. Clearly, this remarkable adaptation arose independently in unrelated fishes.

LITERATURE. Carey 1973; Block et al. 2001

FIELD DESCRIPTION. The slender fusiform body (i.e., compressed, torpedo shape that tapers more at the tail than at the head) is typical of tunas, although this species has some distinct qualities. In large specimens, the dorsal and anal fins are unusually long, over 20% of fish's fork length. Body length measured in scombrids is commonly the length from the tip of the snout to the center of the "fork" in the lunate tail (fork length). For fishes that do not have forked tails, the measurement is from the snout to the end of the most posterior point on the caudal fin (total length).

The dorsal skin is dark blue, the ventral is silvery gray, and the sides are golden. Most of the fins are a bright yellow and there is some yellow on the upper sides. On the lower sides, stripes appear alternatively in an unbroken line and in a chain of dots.

ECOLOGY/LIFE HISTORY. In contrast to many tunas that are open ocean pelagics, yellowfins are migratory coastal pelagics. These tunas, bigeye and skipjack are occasionally caught together during commercial fishing operations. In particular, yellowfin tunas often swim beneath porpoises leading to the unintentional injuring of the porpoises.

FISHERIES AND MANAGEMENT. See the sections below on Tuna Fisheries and Management of Tunas, Billfishes and Swordfish for details on fisheries and management of this species.

Bigeye tuna

Thunnus obesus (patudo)

Worldwide tropical and subtropical
Max. length = 6.3 ft.

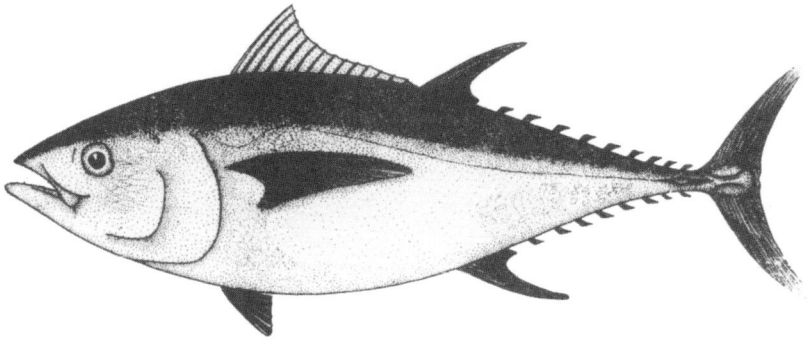

FIELD CHARACTERISTICS. Stocky body and big eyes characterize this species. The pectoral fin in adults extends beneath the second dorsal fin. The dorsal surface is metallic dark blue and the ventral is silvery white. The sides are yellowish purple.

ECOLOGY/LIFE HISTORY. Bigeye and other tunas swim long distances and swim continuously. During that time, they may be in warm surface waters and then dive to frigid waters. Muscle contractions are slow and less powerful at cold temperatures. In response to this challenge, these fishes can physiologically maintain warm parts of their body by the use of a counter current heat-exchanging network of blood vessels. Metabolically heated warm blood from the interior is conserved by warming the cooler blood flowing to the interior.

FISHERIES AND MANAGEMENT. See the sections below on Tuna Fisheries and Management of Tunas, Billfishes, and Swordfish for details on fisheries and management of this species.

Bluefin tuna

Thunnus thynnus (atun aleta azul)

Other common names: giant tuna, horse mackerel
In the western Atlantic from Labrador to Brazil, in the east from Norway to the
        Canary Islands, and Mediterranean
Max. length = 9.8 ft, common length = 6.5 ft.

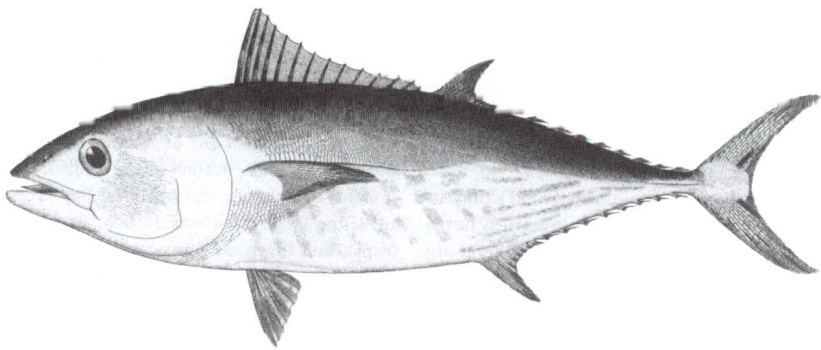

FIELD CHARACTERISTICS. A fusiform body is compressed and very robust in
front. Short pectoral fins do not reach beyond the base of the second dorsal
fin. The height of the second dorsal is longer than that of the first dorsal.
Dorsal skin is dark blue or black without any lines. Ventral skin is silvery gray
with colorless transverse lines and rows of colorless dots. The lateral caudal
keel is black in adults.

ECOLOGY/LIFE HISTORY. This is the largest species within this family. An average
mature female can produce 10 million eggs during the spring and summer
spawning season. Of course, only a fraction of those eggs will be successfully
fertilized and lead to viable young. Growth is very rapid, and it has been
estimated that from the time this tuna hatches as a 3 mm larva and when it
reaches full size, it increases in weight by one billion times.

Many tunas, like the bluefin, can never stop swimming. It has been
estimated that they have to move a distance equal to their body length every
second in order to facilitate gill ventilation.

They can swim continuously at 4.6 mph for many days and thereby are
capable of crossing the Atlantic in less than 60 days. While chasing prey their
top speeds are about 45 mph. Tagged bluefins off Florida were recaptured in
Norway, 6000 mi. away.

Bluefin tuna are highly migratory, moving to warmer waters to spawn
in the spring and back to temperate water to feed in the fall. For over 20
years bluefin migrations have been monitored by using smart electronic tags

(pop-up satellite archival tags and archival tags implanted in the fish and recovered when the fish is captured). These tags report daily location, dive depths, and water and body temperatures. After about 6 weeks of attachment, the pop-up tags are released to the surface, and these data are downloaded to a computer via a satellite. Some implanted tags remained for well over a year. Over 377 bluefins were tagged off North Carolina. The tunas maintained a stable core body temperature of 79 °F (26 °C) even though they occasionally made deep (3000 ft. maximum) dives to very cold (37 °F, 3 °C) water. Four distinct migratory patterns were recorded. One group simply moved along our East Coast, a second migrated to the Gulf of Mexico spawning grounds, a third migrated across the Atlantic, and a fourth migrated to the eastern Atlantic and to spawning grounds in the Mediterranean. Finally, these patterns provide evidence of the close relationship between eastern and western populations of bluefins and the need to reduce overfishing in the eastern Atlantic.

LITERATURE. Joseph et al. 1979; Collette and Nauen 1983; Smith and Hasbrouck 1988; Shadwick et al. 1999; Magnuson et al. 2001; Blake et al. 2005

FISHERIES AND MANAGEMENT. See the sections below on Tuna Fisheries and Management of Tunas, Billfishes, and Swordfish for details on fisheries and management of this species.

TUNA FISHERIES

The fisheries for bluefin and BAYS tunas (bigeye, longfin albacore, yellowfin, and skipjack) as well as little tunny and bonito are combined here as they are similar and these species are generally pursued together by New York fishermen. They are also generally managed as a group (with individual species-specific regulations) by NMFS (see Tunas, Billfishes, and Swordfish Management section below).

Bluefin tuna (BFT) is the most highly priced, prized, and valued tuna for both commercial and recreational fishermen. It is in high demand both in U.S. markets and worldwide, particularly in Japan. Most bluefin tuna, as well as many bigeye and yellowfin, landed for commercial sale in the United States are exported. Bluefin tuna, and high-quality bigeye and yellowfin, are typically sold for sushi and sashimi. Ahi tuna served in U.S. restaurants is usually yellowfin or bigeye tuna. Since bluefin is in such high worldwide demand it provides a high return to fishermen, wholesalers, and exporters. The highest record paid for a single bluefin tuna was US$3.1 million in 2019 for a 612 lb. fish at the wholesale market in Japan. However, that was a closely related Pacific bluefin tuna. Atlantic bluefin typically sell for $200–400 per pound in the Japanese wholesale market.

The commercial fishery for tunas in New York greatly expanded in the early 1980s and continued at a robust level through the end of the century. Tuna landings in New York during this time period (and since) were primarily bluefin, bigeye, albacore, and yellowfin with very little skipjack. The rapid expansion of the fishery in New York was based on several factors, including increasing domestic and export demand for BFT and BAYS tunas; reduced cost and increased service of international air freight out of New York area airports that facilitated the movement of fresh fish to other parts of the world (primarily Japan); a proliferation of dockside wholesale buyers with export experience and business ties to foreign fish companies; and the large availability of the various tunas relatively close to New York ports. These factors led to the rapid expansion of the U.S. East Coast longline fleet that followed the coastal tuna migration and fished for several months of the year out of New York ports and to the significant expansion of the rod and reel fishery that targeted tunas for commercial sale. In New York commercial bluefin landings peaked in 1985 at 288,000 lb., but the maximum landed value of $733,500 was in 1987. During the 1980s bluefin landings averaged 100,000–200,000 lb./yr. Landings dropped off in the 1990s to less than 50,000 lb./yr. and have been at these low levels since then.

Bigeye tuna commercial landings peaked in New York at over 750,000 lb. in both 1994 and 1995. Otherwise during the 1980s and 1990s landings averaged between 300,000 and 500,000 lb. Landings have dropped off since 2003 to generally less than 50,000 lb./yr.—an order of magnitude less than during the height of the fishery.

Albacore commercial landings in New York peaked at around 400,000 lb. in 1994 and in 1995. Otherwise during the 1980s and 1990s landings fluctuated significantly and generally ranged between 50,000 and 200,000 lb. Landings have dropped off since 2003 and have been around 50,000 lb./yr. or less.

Yellowfin commercial landings peaked in New York at just over one million pounds in 1987. Landings fluctuated significantly during the 1980s and 1990s and generally ranged between 200,000 and 700,000 lb. Landings have dropped off since 2003 and have generally been 50,000–70,000 lb./yr.

Skipjack tunas, the fourth of the regulated BAYS tunas, have never been important to New York's commercial fisheries, and landings have been negligible. The commercial fisheries for little tunny and bonito have likewise been of very minor importance to New York commercial fishermen. Landings of these two species are generally less than 30,000 lb. each per year. During a few years in the 1990s landings of each approached 100,000 lb., caught primarily by gill-nets and pound nets. Tuna landed in New York is primarily consumed fresh or exported and does not end up being canned. However, canned tuna is the sec-ond most popular seafood product in the United States after shrimp. Most of all canned and pouch tuna is skipjack or albacore.

The primary gears for the commercial harvest BFT and BAYS tunas are longlines and rod and reel. A small harpoon fishery existed in New York for BFT and yellowfin tuna, but that gear is no longer used in New York for these species although it is still popular in some New England states. There were also short-lived drift gillnet and pelagic pair trawl fisheries in New York in the 1980s for BAYS tunas. However, these lasted only a couple of years before these gears were disallowed in the tuna fishery.

All tuna species landings dropped off significantly in New York after 2003. This reduction had several causes. The elimination of the above-mentioned pair trawl and gillnet fisheries had an effect. A large impact was the development of fishery management plans for BFT and BAYS tunas implemented in the late 1990s that instituted management measures to constrain fishing effort and sus-tainably manage the resources. Also, in the late 1980s the FMP for swordfish was developed, which severely limited the longline fishery, and many vessels left that fishery. Since tunas were also a major catch of the swordfish longline fleet, tuna landings were reduced with the size of that fleet.

Tunas are also an important catch in the New York recreational fishery. All tuna species put up a very strong fight when caught on rod and reel and are an exciting catch. The bigger the fish, the stronger and more exciting the catch! Bluefin tuna, for example, can exceed 10 ft. in length and 1,000 lb. Tunas are a desired recreational fish and can test the ability and experience of anglers and offer a real challenge to catch. They are also a highly desired delicacy. Bluefin and BAYS tunas are typically caught offshore, while little tunny and bonito can also be caught inshore.

Although many bluefin caught by anglers are sold (by those who have a per-mit to do so), there are still bluefin that are caught recreationally. Recreational catches of bluefin in New York were high during the mid-1980s through the mid-1990s, with annual landings in the millions of pounds, and then dropped off after that. Landings were highest in New York in 1991 at 5.1 million pounds. Landings have fluctuated widely since 2000 and have ranged from zero pounds reported in 2006 to 1.5 million pounds in 2019. In 2021 landings were down again to 350,000 lb.

Recreational harvest of bigeye tuna fluctuates widely in New York and has been reported as high as 2.8 million pounds in 1986 and as low as zero many years. Landings in 2013–2020 ranged from 95,000 to 2.6 million pounds. Catches of albacore have likewise fluctuated widely. Eight million pounds were caught in 1987 and catches in 2013–2020 (latest year available) have been zero to 1.7 million pounds.

Yellowfin tuna recreational catches in New York have also fluctuated widely. Recreational harvest reached a peak of 55.3 million pounds in 1987 and ranged in the millions of pounds during most of the late 1980s through the 2000s. Harvest in 2013 to 2021 ranged from 19,200 lb. to 4.1 million pounds. Skipjack tuna has never been very important to the New York recreational fishery. Recreational harvest peaked at 952,000 lb. in 1985 but has generally been less than 50,000 lb. There were no reported catches in 2014 through 2021. Little tunny and bonito have likewise been of low importance for recreational anglers as a food fish. However, they are popular for catch and release. For many years there were no reported recreational landings of either species, but landings were occasionally around 10,000 lb. However, peak recreational landings of little tunny occurred in 2019 at 250,000 lb. and in 1992 for bonito at 775,000 lb. As mentioned above, little tunny can swim at very high speeds for long periods of time. This species is therefore extremely fun (and difficult) to catch on rod and reel, so there is an extensive recreational fishery for "albies," but it is primarily catch and release, which is not recorded as recreational landings.

RECREATIONAL WORLD AND NEW YORK RECORDS FOR TUNAS

SPECIES AND LOCATION FOR WORLD RECORD	ALL TACKLE WORLD RECORD (LB.)	NEW YORK STATE RECORD (LB.)
Bluefin (Atlantic Ocean)—Nova Scotia	1496.00	1071.00
Bigeye (Atlantic Ocean)—Puerto Rico	392.375	355.00
Albacore—Canary Islands	88.125	81.15
Yellowfin—Mexico	427.00	248.50
Skipjack—Mexico	45.25	N/A
Little tunny—New Jersey	36.00	16.38
Bonito—Azores	18.25	12.49

MANAGEMENT OF TUNAS, BILLFISHES, AND SWORDFISH

The various species of tunas, billfishes, and swordfish live, and are fished, throughout the Atlantic Ocean and migrate great distances. Because these species cross domestic and international boundaries they are outside the purview of the ASMFC, the MAFMC, and the NEFMC. As such they are managed in the United States by the Atlantic Highly Migratory Species Management Division (HMS) of the NMFS. The sustainable management of these species also requires international cooperation and action. Thus, these species are also managed through the International Convention for the Conservation of Atlantic Tunas (ICCAT), the Food and Agriculture Organization of the United Nations, and the Atlantic Tunas Convention Act. The ICCAT develops international management measures for billfishes and swordfish as well as tunas. Tunas, billfishes, and swordfish are caught primarily in federal waters but occasionally are found in state waters as well. Federal regulations also apply in New York State waters. A federal permit is required to catch all tunas, billfishes, and swordfish by any gear type for either recreational or commercial purposes.

The species of tunas regulated by HMS are bluefin tuna as well as BAYS tunas: bigeye, longfin albacore, yellowfin, and skipjack. Tunas are managed through various permit categories. These categories are general, harpoon, purse seine, angling, charter/head boat, longline, and trap (pound net). Each category has its own set of species-specific regulations and qualifications for participation. Regulations include minimum size, bag or trip limits, and seasons. All vessels (commercial, private recreational, and chart/head boat) fishing for bluefin or BAYS tunas must have a permit. Atlantic tunas may be sold only by fishermen permitted in commercial categories and only to permitted dealers. In New York, any fisherman selling tunas must also have a New York food fish or landing (caught outside New York waters) permit.

Bluefin tuna is the most heavily regulated of the HMS species under a very complex system of quotas, subquotas, regions, possession limits, and size limits for each of the seven permit categories mentioned above (NMFS 2006). Bluefin is further submanaged by six different size categories. Longline and purse seine vessels also have an individual bluefin quota for each vessel. Longline vessels are further required to have installed and approved electronic monitoring equipment to video record all their catch. They are also required to carry federal observers. All recreational catches and discards of bluefin must be reported.

For management purposes billfishes include sailfish, blue marlin, longbill spearfish, and white marlin. Longbill spearfish is prohibited for recreational catch. The other billfish species can be caught recreationally with an angling or charter/head boat permit. Tuna general category or swordfish general commercial permittees can also fish in registered HMS tournaments. Recreational regulations also include species minimum sizes, hook type allowed, and an overall combined coast-wide quota. All recreational catches of billfishes must be reported. The sale of billfishes is prohibited by all gear types, so there is no longer a commercial fishery for billfishes.

Swordfish is managed for the recreational fishery similar to the recreational management of billfishes, including permit categories as described above. Minimum sizes and bag limits apply. The commercial swordfish fishery is managed through an overall quota, trip limits, and permit categories. There are six permit categories as follows. The directed permit and handgear permit are limited entry and are for those fishermen directing their effort primarily on swordfish. These two permit categories allow for the greatest swordfish harvest. There are also the incidental permit (also limited entry), swordfish general commercial permit, charter/head boat permit, and incidental squid trawl permit. These various permit categories have various restricted trip limits and gear restrictions.

SWORDFISH
Family Xiphiidae (*espadas*)

This family contains only one species. As tempting as it may appear, swordfish and billfishes (marlin, sailfish, and spearfish) have never been placed in the same family despite some conspicuous common features, for example, the rostrum produced from an elongated premaxillary bone, an inferior mouth, the pectoral fins placed low on the body, two anal fins, and a large lunate caudal fin designed for maximum thrust. In swordfish, the flattened and smooth bill appears very sword-like. The single species in the family has a scientific name that refers to the sword twice: the genus and species in Greek and Latin for "sword" respectively.

Swordfish

Xiphias gladius (pez espada)

Other common names: broadbill
In all temperate and tropical oceans. In North America from Newfoundland to the
Gulf of Mexico.
Max. length = 17 ft., weight = 1400 lb.

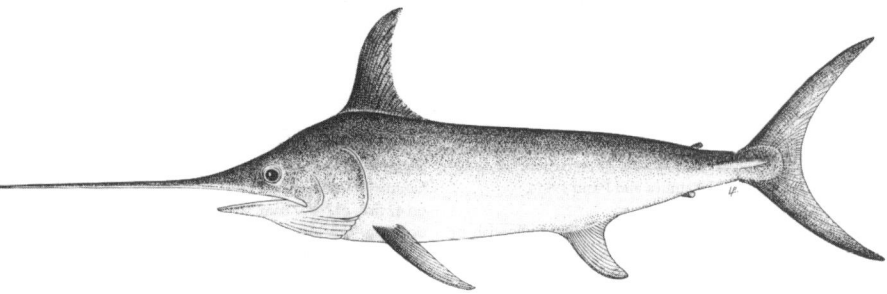

FIELD CHARACTERISTICS. In contrast to the billfish family, the first dorsal fin
has a short base, the caudal peduncle has single lateral keels on each side, the
pelvic fin is absent, and so are any teeth and scales in adults. Larval swordfish
have well-developed ctenoid scales that confer protection, but as the fish
matures the scales become deeply embedded in the dermis.

The swordfish is strongly counter-shaded with dark black to dark blue
dorsal surfaces and a lighter ventral area. The sides have a coppery hue. This
coloring is generally true of billfishes as well, although swordfish do not have a
series of bars along the side.

ECOLOGY/LIFE HISTORY. The sword is the most prominent feature of this fish
and has been observed to be used to slash at prey (schools of squid and
fishes), stunning and/or slicing them. Consider that the sword's length can be
one-third of the fish's total body length—some swords may reach over 4 ft.
long. Large adult females may weigh over 1000 lb. and have bursts of speed
estimated to be close to 60 mph. The force they can generate is formidable.
The ichthyologist George Brown Goode reported an incident in which a
sword was discovered that had penetrated the hull of an 1828 whaling ship.
The hull consisted of copper sheathing and 16 in. of hardwood, after which
the sword penetrated a thick ceiling and an oak cask.

The swordfish begins life as a larval fish with a long snout (upper
jaw premaxillary) and a lower jaw (mandible) of the same length. The
disproportionate development of the upper and lower jaws results primarily
from the retarded growth of the mandible rather than the accelerated growth
of the snout. The functional significance of this is obvious in that it produces

WARM HEADS

Swordfish possess a "brain heater." This is a thermogenic structure that has been modified from an eye muscle. This structure locally warms brain and eye tissue and thus contributes to greater foraging success in cold, deep waters. Such an adaptation has also been developed in billfishes and tunas. An exceptional example of whole body endothermy, that is, the metabolic production and retention of heat, has been reported in the opah (*Lampris guttatus*), a pelagic fish that inhabits the cold, deep waters from Nova Scotia to South America. (This offshore species is not included in a listing of New York marine fishes.) The heat is generated by the continuous flapping of the opah's pectoral fins. That heat is retained by gill heat exchangers and is distributed by the heart throughout the body.

LITERATURE. Goode 1884; Carey and Robison 1981; Carey 1982; McGowan 1988; Govoni et al. 2004; Wegner et al. 2015

the long sword that is an effective instrument for slashing prey. The sword, however, is not absolutely critical for feeding success. Instances in which swordfish have broken-off swords appear well fed. Another possible advantage in having a shortened mandible is that the mouth cannot be irreversibly and fatally closed by objects accidentally impaled on the sword.

The swordfish is oceanic and has been monitored performing daily upward migrations from great depths (1500 ft.) at dusk, which are seemingly triggered by a falling light intensity. At dawn, as anticipated, they move downward again.

FISHERIES. Swordfish became important to the New York commercial fishery during the early 1930s and 1940s when landings ranged from 50,000 lb. to 300,000 lb. per year. Landings then dropped off to low levels. There was a slight resurgence in the late 1950s and early 1960s when landings ranged from 50,000 lb. to 170,000 lb. annually. During this time, and in the 1930s and 1940s, swordfish were harvested by harpoon. Landings experienced a resurgence and expansion during the 1980s through 2011 with the development and expansion of the longline fishery. Landings quickly rose to 500,000–600,000 lb./yr. when New York longline vessels as well as longline vessels from southern states followed the fish north and landed them in New York. In the 2000s landings fell to 100,000–200,000 lb./yr. as management regulations began to reduce landing and effort in the longline fishery. Since 2014 landings have been around only 50,000 lb./yr. Peak landings occurred in 1986 at 617,000 lb.

The recreational fishery in New York for swordfish has been very sporadic although it is an exciting fish to catch on rod and reel. Fishermen with the boats and gear to pursue this fish, usually offshore, tend to target catching tunas. Recreational landing estimates can be sporadic or imprecise, particularly for species like swordfish for which there may not be many fishermen or a lot of trips catching swordfish. For the few years that there are reported recreational swordfish landings, landings were 170,000–200,000 lb. for only a couple of years from the 2010s into the 2020s. Most years during this time period had no reported catch.

ALL TACKLE WORLD RECORD: 1182 lb., Chile

NEW YORK RECORD: 540 lb.

MANAGEMENT. See the section titled Management of Tunas, Billfishes, and Swordfish.

BILLFISHES
Family Istiophoridae (*picudos*)

Five species occur along the U.S. Atlantic coast: blue and white marlins, round-scale and longbill spearfishes, and the sailfish. An additional six other species constitute this family, which has a worldwide presence in tropical and subtropical oceans.

White marlin

Kajikia albida (marlin blanco)

Nova Scotia to Florida, Gulf of Mexico to Brazil, less common in the north
Max. length = 9.1 ft., common = 5.5 ft.

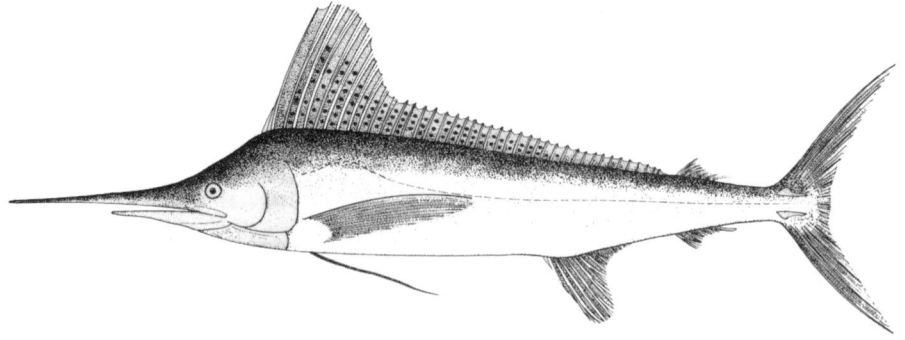

TOXIC FISHES

Not too many years ago, smoked marlin salad sandwiches on pumpernickel were a New York favorite. However, marlin is no longer available as a commercial fishery. Fishermen must release all marlins within 200 mi. of the coastline. If legal marlins are landed recreationally, they have to be at least 8.25 ft. in length.

Further, vendors must assure that scombroids (tuna, mackerel, billfishes, and swordfish) are adequately preserved and refrigerated. If not, scombroid poisoning, also called histamine poisoning, may occur. Apparently, certain fishes (scombroids but also bluefish, dolphinfish, jacks, herring, and anchovy) contain high levels of histidine that can be converted to histamine by bacteria. This results in severe allergic reactions, including flushing of the skin, nausea, and headaches.

This latter fish "poison" is the result of mishandling the fish rather than from a lower trophic level source as is the case for ciguatoxin. That toxin is produced by warm-water dinoflagellates fed upon by herbivores and ultimately biologically magnified through the food chain and concentrated in large carnivorous tropical and subtropical fishes. Ciguatera, as the poisoning is called, is considerably more dangerous to humans than histamine poisoning, but ciguatera poisoning should not occur in fishes harvested from New York waters.

LITERATURE. Hungerford 2010

FIELD CHARACTERISTICS. In contrast to the blue marlin, the white marlin's pectoral fin tips are rounded and those fins are shorter than the pelvic fins. Further, the anterior portion of the dorsal fin and the tip of the anal fin are also rounded.

Blue marlin

Makaira nigricans (marlin azul)

Both sides of the Atlantic Ocean in temperate and tropical waters
Max. length = 14 ft.

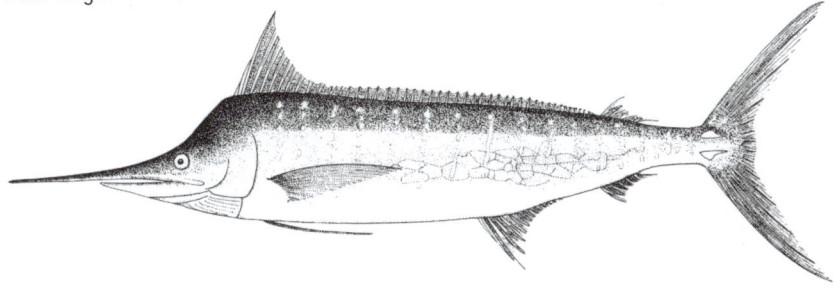

FIELD CHARACTERISTICS. The most prominent feature is the upper jaw, which is extended out to a round spear-like bill. Unlike the swordfish, the billfishes have a long-based first dorsal fin. That fin fits into a groove when the fish is swimming fast. Further distinctions include long and narrow pelvic fins, a caudal fin with two keels at the base, and adults who possess teeth and scales.

The blue marlin has a dark blue dorsal surface, several vertical bars on its side, and pelvic fins that are shorter than the pectorals. The other members of this family vary in the length of their pelvic fins and the nature and color of the bars. The blue marlin is the largest of the billfishes.

ECOLOGY/LIFE HISTORY. The blue marlin, like most other billfishes, tends to be a solitary pelagic fish. Although blue and black marlins can achieve bursts of very high speed (46 and 80 mph, respectively), sustained swimming speeds including short spurts may be only between 2 and 5 mph. As adults, blue marlins undergo yearly transoceanic migrations, swimming slowly to minimize energy costs. Among the features that enhance its swimming ability are the marlin's elongate bill, streamlined body, pectoral fins that provide lift, and the strongly lunate tail that efficiently produces thrust. A large proportion of their muscle mass is red muscle. Red muscle plays an important role in sustained swimming. That muscle is highly vascularized and contains oxygen-rich myoglobin and an abundant supply of energy-producing mitochondria. These muscles are less prone to fatigue.

Marlins feed on epipelagic fishes and squid by using their bill to stun prey. The blue marlin is a legendary fighting game fish. Hemingway's fictional giant fish in *The Old Man and the Sea* was a blue marlin.

LITERATURE. Block et al. 1992

BUTTERFISHES
Family Stromateidae

There are three American Atlantic species, but only one occurs off Long Island. Fifteen other representatives of the family occur throughout the world. Many of these species, as juveniles, find shelter beneath floating objects (seaweed, jellyfish). The man-of-war fish, a tropical member of a closely related family, acquires its common name from its association with the venomous Portuguese man-of-war, a planktonic jellyfish.

In general, butterfishes have no pelvic fins, a forked caudal fin, and small mouths.

Butterfish

Peprilus triacanthus

Other common names: dollarfish
Nova Scotia to South Carolina
Max. length = 12 in.

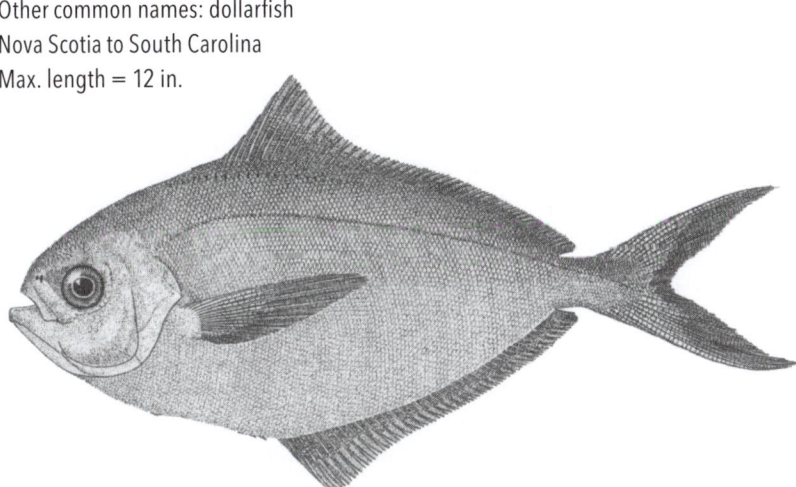

FIELD CHARACTERISTICS. It is assumed that the species name *triacanthus* refers to the number of small, often very inconspicuous, short spines that precede the anal and the dorsal fins. A curious feature of these dorsal and anal spines is that the first one points forward. The butterfish possesses all the standard family features as well as relatively large eye.

In general, live fishes are difficult to handle. The butterfish is no exception and may even be more so. Its body is deep and laterally compressed and the skin has a generous coat of mucus. Further, the small cycloid scales rub off easily.

The butterfish has a simple, slightly arched lateral line extending from its gill cover to the caudal peduncle. This lateral line runs relatively high on the body, which is often the case for fishes that commonly live near the bottom as the butterfish adult does. In addition to the small lateral line pores there are 17–25 larger pores under the anterior half of the dorsal fin. The function of the lateral line is well understood as a distance-touch receptor, but the function of the larger subdorsal fin pores is unclear. Those latter pores may have a sensory function or act as a subdermal mucosal system.

ECOLOGY/LIFE HISTORY. A copious supply of skin mucus may contribute to the relative immunity these fish have if they, as juveniles, come in contact with their potent commensal hosts. As adults, butterfish prefer sandy bottoms, but at greater depths they may be near the surface; inshore they occur nearer to the bottom.

The diet includes ctenophores and other soft-bodied invertebrates as well as small crustaceans, for example, copepods, amphipods, and euphausid and mysid shrimp. The digestive tract of butterfish appears to be specialized in the ability to process the ingested crustaceans. There is a small muscular sac located in an expanded portion of the esophagus. The sac is lined by mucous glands and many small "teeth."

LITERATURE. Dijikgraaf 1962; McDowall 1981

FISHERIES. Butterfish is caught primarily by commercial fishermen. There is essentially no recreational fishery for them since they feed on plankton and will not bite a baited hook or artificial lure. Although prized as a food fish in other parts of the world, particularly when their fat content is high, their consumption in the United States is low. However, there is strong foreign demand, and much of the domestic landings are exported. Butterfish are a favorite prey of many gamefish, including bluefish, striped bass, tunas, and other pelagic fishes. Thus, butterfish is used frequently as bait in recreational fisheries so some of the commercial landings are sold as bait. The principal commercial fishing gear for butterfish is the otter trawl, and they are often caught while fishermen are pursuing other species (primarily squid). Butterfish is also caught in pound nets. New York landings of butterfish peaked in the late 1930s shortly after the broad adoption of the otter trawl, at around five million pounds. Landings since the early 1970s have fluctuated between a quarter million and one million pounds annually in New York.

MANAGEMENT. Butterfish is managed by the MAFMC and NMFS in federal waters as part of the Atlantic Mackerel, Squid, and Butterfish FMP. The federal waters (outside 3 mi.) management measures incorporate trawl mesh size and quotas, including a bycatch quota for the squid fishery. If the bycatch quota is exceeded, then the squid fishery can be closed down. Squid fishermen thus participate in an avoidance communication network (squidtrawlnetwork .com) to avoid areas where there are large concentrations of butterfish. New York does not have any management measures on butterfish within New York waters (within 3 mi.).

Butterfish is short-lived, fast growing, and eaten in large numbers by other species and thus can experience wide fluctuations in abundance. Although previously overfished, at the time of publication of this book, it is no longer overfished and overfishing is not occurring.

TURBOTS
Family Scophthalmidae (*rodaballos*)

The large order of flatfishes includes 678 species classified within 14 families. Six of those families occur to varying extents along the Atlantic coast of the United Stats. Although they all share the compressed body with asymmetrically placed eyes and pigmentation, they are all distinguished externally by whether they are right or left eyed, have straight or arched lateral lines over their pectoral fins, the length of upper jaws and size of the mouth, similarities of pelvic fins, the presence of pectoral fins, and scale type. The process whereby the eye on the underside completely migrates around the top of the head is accompanied by a reorganization of bones, muscles, and nerves. It is temperature dependent but generally this remarkable metamorphosis occurs within the first 6–8 weeks of development.

There are eight species within this left-eyed flatfish family, but only one occurs in the western Atlantic.

Windowpane

Scophthalmus aquosus

Other common names: sundial, sand dab, brill
Gulf of St. Lawrence to Florida, not common north of Massachusetts
Max. length = 20 in.

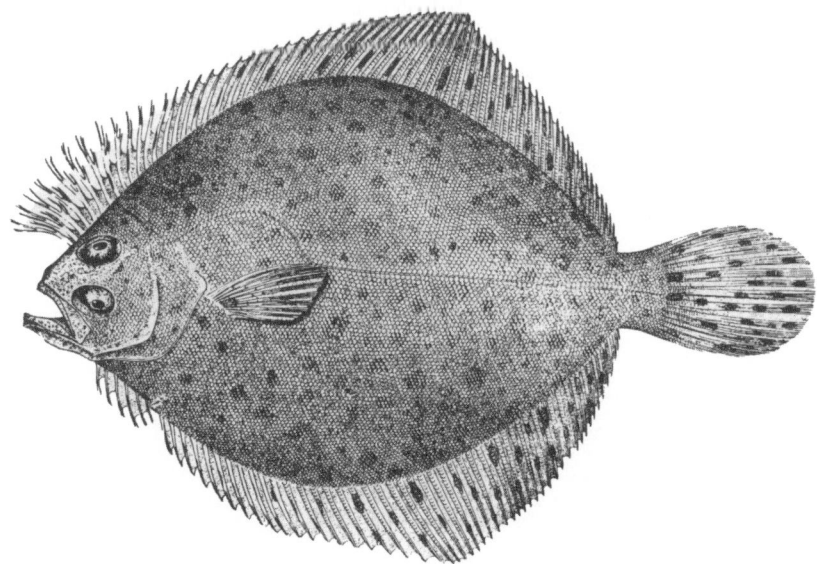

FIELD CHARACTERISTICS. Thin, nearly translucent body especially when held up against a sunny sky. Most anterior dorsal fin rays (10–12) are free, followed by the remaining 52–60 soft rays of the single dorsal fin. Mouth is fairly large but teeth are small. The pelvic fins are placed just below the level of the eyes, followed immediately by a long anal fin that continues to the caudal peduncle. The body with extended fins is rhomboid in shape. Scales are smooth (cycloid) on both the eyed and blind sides. The lateral line is highly arched over the pectoral fin.

Windowpane do not support a substantial recreational or commercial fishery in New York.

SAND FLOUNDERS
Family Paralichthyidae (*lenguados areneros*)

The sand flounder family is the largest flatfish family (105 species) and occurs worldwide in tropical and temperate seas. It is left eyed, but this family is not the only left eyed family on the East Coast of the United States. The other left-eyed families are the turbot (Scophthalmidae), lefteye flounder (Bothidae), and tonguefish (Cynoglossidae). The latter family is only rarely seen as far north as New York, represented by the blackcheek tonguefish, *Symphurus plagiusa*.

Smallmouth flounder

Etropus microstomus

Most common from Cape Cod to Cape Hatteras, North Carolina, but also to Florida
Max. length = 6 in.

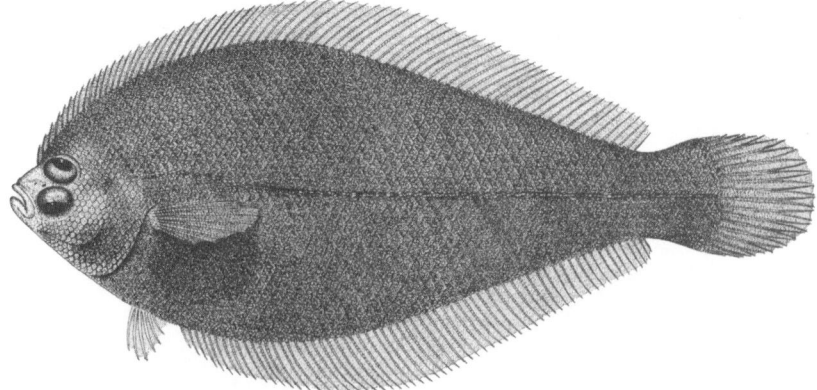

FIELD CHARACTERISTICS. The species name is from the Greek terms for small mouth. The mouth is markedly small and barely reaches the anterior edge of the lower eye. The eyes are close together. The lateral line is relatively straight. The pectoral fin on the eyed side is larger than the one on the blind side. The scales on the eyed side are ctenoid but cycloid on the blind side.

Summer flounder

Paralichthys dentatus

Other common names: fluke, doormat
Nova Scotia to South Carolina
Max. length = 2.7 ft.

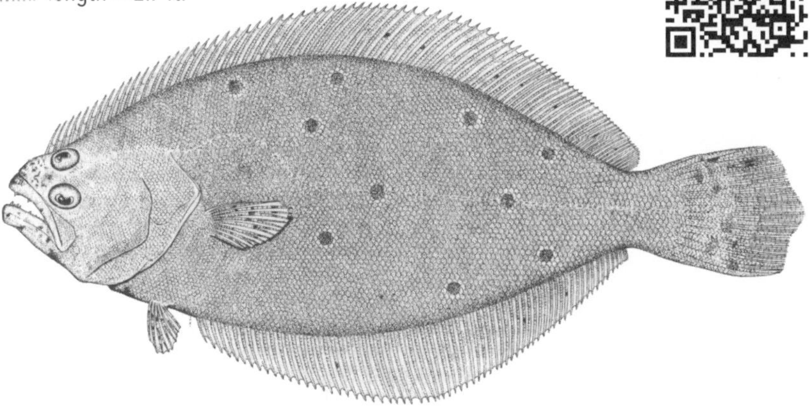

FIELD CHARACTERISTICS. The large upper jaw ends at the rear margin of the eyes and the canine teeth are formidable in this *dentatus* species. Other qualities include small cycloid scales, an arched lateral line over the pectorals, and a rounded caudal fin. The color is variable and can be light or dark shades of brown. Often there are five large occeli (eye spots) and numerous smaller spots.

ECOLOGY/LIFE HISTORY. The official common name "summer flounder" accurately celebrates the seasonal appearance of this fish during the warm months. At that time, they can be found over inshore sandy and mud bottoms or at modest depths of 30–40 ft. During the fall they move offshore to the continental shelf and depths of 120 to 600 ft. and spawn in the fall and winter. A large female may produce over three million eggs.

As is seen in other fishes, females grow faster and larger than males. One can assume, that the other common names "fluke" and "doormat" refer to large specimens that resemble the lobe of a whale's tail or a door mat.

Summer flounder are more active predators than other more lethargic benthic flatfishes. These fish have been observed to aggressively pursue prey. That prey might include young winter flounder, menhaden, weakfish, sand lance, blue claw crabs, squid, and sand shrimp.

LITERATURE. Bolker et al. 2005; Cernadas-Martin et al. 2021

FISHERIES. Summer flounder has historically been very important to New York's commercial and recreational fisheries and continue to be so. There is high market demand for this flavorful flounder. They are also kept alive or bled and handled in a specific manner for the sushi market. They bite readily on a baited hook, put up a good fight, and grow to a large size for a flounder.

Summer flounder is one of the most sought after commercial and recreational fish along the Mid-Atlantic and southern New England coast. Fishing areas change with the migrational patterns of summer flounder: an inshore recreational and commercial fishery in spring, summer, and fall and an offshore deep-water commercial fishery in winter.

Summer flounder commercial landings in New York increased during the 1940s and 1950s, reaching an historical maximum for New York of over four million pounds in 1956. The catch declined moderately in the late 1960s and early 1970s. Landings ranged between 1.5 and 2.5 million pounds during the 1980s, then reduced to less than one million pounds in the 1990s as the stock of summer flounder declined and quotas were implemented. Commercial landings are limited by state-by-state quota and since 2010 have been around 600,000 lb. to 1.5 million pounds for New York. Otter trawls account for most of the commercial landings in New York. This flounder is also caught commercially by hook and line and pound net.

Summer flounder is highly sought by recreational fishermen in New York and throughout the Mid-Atlantic. The East Coast recreational harvest peaked in 1983 when 37 million pounds were caught. Since 2018 the East Coast recreational harvest has averaged around eight million pounds annually. In New York the recreational harvest has averaged around four to five million pounds and is limited by a regional recreational harvest limit. Most recent recreational harvest has been around 2.5 million pounds. New York catches around one-third of the coastal recreational catch.

ALL TACKLE WORLD RECORD: 22 lbs. 7oz., New York

NEW YORK RECORD: 22 lbs. 7oz.

MANAGEMENT. Summer flounder is managed jointly by the MAFMC, ASMFC, NMFS, and the NYSDEC though the Summer Flounder, Scup, and Black Sea Bass FMP (MAFMC 1987). Summer flounder is sustainably managed and

COLOR CHANGE

Although most flatfishes are capable of assuming the shade and, for some, the pattern of their background, the summer flounder is recognized as being one of the more chameleon-like. Most commonly, when not buried in the sandy substrate, this species is nearly indistinguishable from the sandy bottom upon which it lives. Color change in fishes is complex in terms of the pigment cells involved and the factors that influence them. Pigment cells (chromatophores) located in the skin contain pigment packets of black melanin (melanophores) and of red or yellow carotenoid (erythorphores and xanthophores). Other chromatophore inclusions contribute to the variety of colors fish can assume. The distribution of the pigment packets within the cells determines whether these epidermal cells will appear dark or light. For example, if the melanin pigment is concentrated in the center of the cells, the body will appear pale. If the pigment is dispersed the body will darken. Neural and hormonal factors cause the changes in pigment distribution. These changes can be rapid and short term or slow and long lasting. Typically, light receptors in the fish's eyes and elsewhere evaluate the light impinging on the bodies from above and that which is reflected off the bottom. Ultimately, nerves that are associated with the pigment cells will release neurotransmitters that affect the state of the cells.

was declared "rebuilt" in 2012 after many years of being considered depleted. As of the publication date of this book, the summer flounder stock is neither overfished nor is it experiencing overfishing (NMFS 2019). The FMP for summer flounder was initiated in 1982, and the first quota was established in 1993. The division of the overall quota was recently revised and allocated 55% commercial and 45% recreational. State-by-state quota allocations for the commercial and recreational fisheries were established using baseline data from 1980 to 1989. The commercial fishery is still managed on a state-by-state basis. Owing to inequities and inadequacies in how commercial catches in New York were tracked during the baseline period, New York ended up with a disproportionately low share of the quota. (7.65%)

Given its high importance to both commercial and recreational fishermen, summer flounder management has been a very contentious issue since the development of the FMP. New York divides its commercial allocation into seven different time periods, each with its own subquota. The commercial fishery is further regulated by trip limits, minimum fish size, minimum mesh size in trawls, and a moratorium on entry into the fishery.

The recreational fishery is managed regionally to allow for some consistency between adjoining states. New York and Connecticut are in the same region.

Recreational fishery management measures include a combination of minimum size limits, bag limits, and fishing seasons and an overall recreational harvest limit.

In 2022, the ASMFC and MAFMC approved a harvest control rule approach for the recreational fisheries management of summer flounder, scup, black sea bass, and bluefish. The changes include a new process for setting recreational measures (bag, size, and season limits). Approval of this new process is part of a broader long-term effort to improve recreational management of these four species. The new management program aims to provide greater stability and predictability in recreational measures from year to year while accounting for uncertainty in recreational catch estimates.

Fourspot flounder

Paralichthys oblongus

Georges Bank to Florida
Max. length = 17 in.

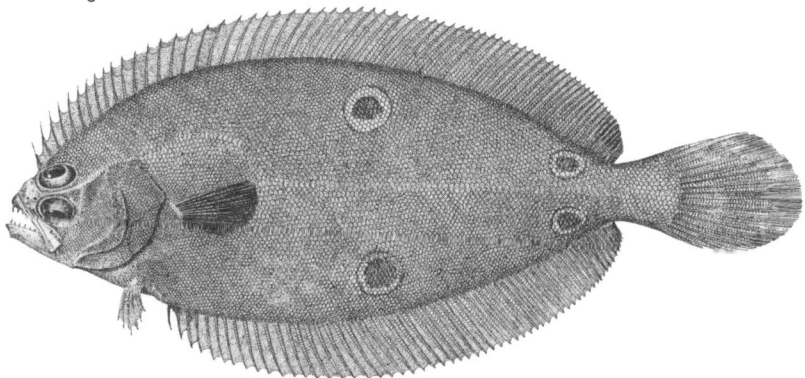

FIELD CHARACTERISTICS. The body is ovate not oblong. The species name *oblongus* refers to the shape of the distinct four eye spots, two of which occur on the midbody and two just in front of the caudal peduncle. The mouth is large and extends back to the posterior margin of the lower eye. The pectoral fin on the eyed and blind side are equal in size. The lateral line is arched over the pectoral fin. The scales on the eyed side are cycloid.

LITERATURE. Mast 1916; Brewster 1987

RIGHT OR LEFT EYED?

How does one know whether a flatfish is right or left eyed? Hold the fish with the pigmented side in full view with the belly of the fish nearest to you. (The pectoral and pelvic fins and lower jaw are all below the midline of the fish so the belly is easily determined). Holding the fish in that position, you will note that the eyes and the mouth of the specimen are pointed to your right or left. Curiously, in any population of flounders there is a small proportion of mutants that have the reverse of the orientation expected.

RIGHTEYE FLOUNDERS
Family Pleuronectidae (*platijas*)

This family includes 23 genera and 60 species, but only 6 species live off the East Coast of North America and 5 of those occur, to varying degrees, in New York waters. This family is one of several flatfish families found in the Northern Hemisphere. Flatfishes develop from bilaterally symmetrical larvae to become either right (dextral) or left (sinistral) eyed. In the first weeks after hatching, flatfish larvae live in the plankton. In the winter flounder, as the larvae grow to about 9 mm, the left eye migrates to its final position on the right side, resulting in both eyes being on that side. This asymmetry applies not only to the eyes but also to the skin, in that the "blind" side is not pigmented. Other less noticeable asymmetries occur in the fish's scale type, skull shape, teeth development, and pectoral fin length. Metamorphosis is generally complete in 2–3 months, during which the fish settles and becomes primarily a benthic creature. Some flounders are adept at blending into the background and, in loose sediment, can bury themselves with only their eyes showing. Pleuronectes, the term applied to the order of flatfishes, means "side swimmer" and refers to the unique mode of locomotion seen in flatfishes.

Yellowtail flounder

Limanda ferruginea

Other common names: rusty dab, yellowtail, rusty flounder
Labrador to Chesapeake Bay
Av. length = 20 in.

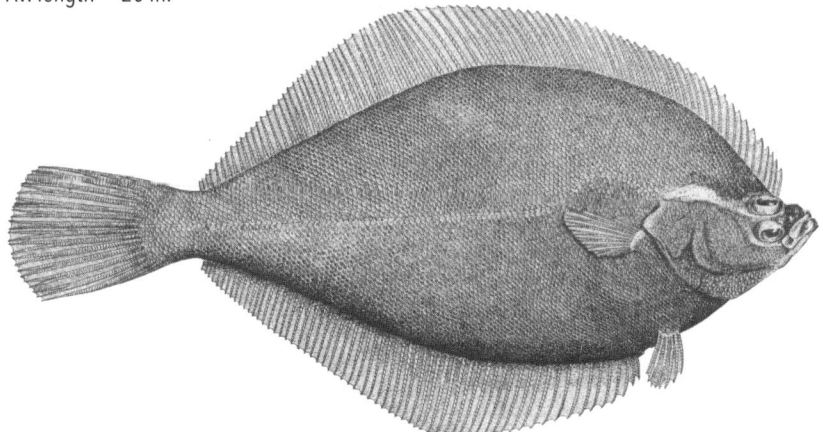

FIELD CHARACTERISTICS. Like all other flatfishes, the body color of *Limanda* is variable, but the eyed side is often brown to olive gray and very routinely has distinctive irregular rusty red spots. The name *ferruginea* refers to those rusty red spots, but the standard common name "yellowtail" arises from the yellow color seen on the tips of the long dorsal and anal fins in addition to the color seen on the margin of the caudal fin.

The body is oval and compressed, the jaws are small, and the eyes are very closely set, separated by a high, narrow ridge. The scales on the eyed side are ctenoid but are cycloid on the blind side. Additional features distinguish this flounder from other relatively small righteye flounders (i.e., American plaice, smooth flounder, witch flounder, and winter flounder). These features include a strongly arched lateral line above the pectoral fin and a head with a concave upper profile.

ECOLOGY/LIFE HISTORY. This species lives offshore on the continental shelf at varying depths but most commonly between 30 to 300 ft. It prefers sandy or muddy surfaces and seems to avoid rocky areas. The genus *Limanda* has its origin from the Latin word *limus* for mud.

This is a particularly sedentary species that appears not to undergo an extensive migration. The small mouth and its sedentary habit influence its diet of relatively small crustaceans, mollusks, and worms.

FISHERIES. Yellowtail flounder is an offshore species that does not come inshore, nor does it put up much of a fight if caught on hook and line. Thus, the fishery in New York is a commercial one, with no recreational catch. Their past abundance and clean sweet-tasting fillets created major fishing pressure on this species by commercial fishermen. However commercial landings have declined significantly since the early 2000s and are at a fraction of their historic levels due to abundance issues, shifting distributions patterns, changes in management, and fishing pressure. Yellowtail flounder historically has been a prime commercial target species because of its easy availability, consumer demand, and, in turn, consistent good market price. In New York during the 1960s and 1970s any "flounder" on the menu or in the fish market was likely yellowtail flounder. The principal commercial fishing gear is the otter trawl. New York landings were high during the World War II era and reached a peak of 11.8 million pounds in 1942. Landings dropped off in the 1950s, then picked up again in the 1960s and 1970s when landings were two to three million pounds per year. New York commercial landings over the past 20 years have been less than a half million pounds and more recently less than 100,000 lb.

MANAGEMENT. Yellowtail flounder is managed jointly by the NEFMC and NMFS as part of the Northeast Multispecies (groundfish) FMP, which also includes a variety of other species including haddock, windowpane, winter flounder, and cod. The fishery is managed as three separate stocks: Georges Bank; Cape Cod/Gulf of Maine; southern New England/Mid-Atlantic. All three stocks are overfished and are subject to overfishing. Fishing is still allowed but at reduced levels. All three stocks have been in a rebuilding mode for many years as part of the FMP. This rebuilding program is based on overall quotas, minimum size, seasons, gear restrictions, trip limits, and individual allocations in the commercial fishery.

Winter flounder

Pseudopleuronectes americanus

Other common names: blackback, lemon sole, snowshoe
Labrador to Georgia, most abundant from St Lawrence to Chesapeake Bay
Max. length = 26 in.

FIELD CHARACTERISTICS. Winter flounder is the most common right-eyed flatfish in New York's shallow bays, especially during the winter spawning period. During the warmer months, these fish tend to move offshore, although there is a part of the population that occurs as year-round resident fish within estuaries and bays. In general, Winter flounder don't move great distances compared with other fishes. Body design features that may promote efficient bottom living are not ideal for long distance migrations. Color and pattern are highly variable and unreliable for identification purposes.

ECOLOGY/LIFE HISTORY. The winter flounder is omnivorous, but its small mouth limits the size of animal prey to amphipods, isopods, shrimp, other small crustaceans, worms, and tips of clam siphons. If a flounder avoids predators and fishermen, it may be about 12 in. in length by its fourth year. Older flounder can get twice that size and are called "snowshoes."

The origin of this flounder's common name is clear. In New York, winter flounder spawn from late December to early May in shallow, often freezing and ice-laden, water. This is an effective antipredator strategy. Other fishes that may be predatory are not active during this time. Those latter fishes are, unlike winter flounder, unable to avoid freezing under those conditions whereas winter flounder blood contains biological antifreeze molecules. (In the summer, these flounder are intolerant of very warm water, and those that do not migrate to deeper, colder water in the summer can somewhat avoid high temperatures by burying themselves in sand cooled by seeping groundwater).

FISH ANTIFREEZE PROTEINS

During the winter, New York seawater is often ice-laden and near its freezing point. This is a problem for marine fishes because these fishes' body fluids (e.g., blood) have a lower concentration of salts and other solutes than found in seawater. As a result, the internal fluids of the fish have a lower freezing point than full seawater (−1.9 vs. −0.8 °C). Under these cold conditions, the fish are supercooled, and if they come in contact with ice, ice crystals will form within the body fluids, resulting in the fish dying. In response to this challenge, about a dozen unrelated fish families have independently evolved genes that direct the synthesis of antifreeze proteins. It is suggested that the history of glaciations in the past 2.5 million years has driven the multiple independent evolution of these adaptive proteins in eight northern families and four families that are found in only the Antarctic and the Southern Ocean. The antifreeze molecules haves been classified into five types, that is, Type I–IV peptides and a glycopeptide. Each family produces only one type, and there are distinct structural differences in the antifreeze found in each species. Despite those differences, the proteins share enough in common so that the antifreeze protects the fish in the same way, that is, by binding to ice crystals that may enter the fish and inhibiting the growth of those crystals. These proteins are primarily synthesized in the liver, but other tissues have been implicated in producing antifreeze glycopeptides in some species.

Common local fishes that breed in the winter in shallow ice-laden water (winter flounder, grubby sculpin, and the Atlantic tomcod) have been shown to produce antifreeze seasonally just before the protein is needed, and it is retained until freezing is no longer a problem.

Freshwater fishes do not have the same problem as marine fishes. In freshwater, fishes' internal fluid salt concentration is much greater than that found in freshwater and so the fish has a lower freezing point and will not freeze before the freshwater does.

LITERATURE. Potzel et al. 1990; Reisman et al. 1987; Fletcher et al. 2001; Cheng and DeVries 2002

The peak for winter flounder spawning is in February and March. Female flounder mature after 2–3 years at a length of about 10 in. Females vary in the number of eggs produced per season. An average female may produce as many as 500,000 eggs but a 16 in., 5-year-old female was reported to contain over three million eggs.

FISHERIES. Winter flounder is targeted by both recreational and commercial fishermen in New York and states to the north. Their past abundance, clean sweet-tasting fillets, and nearshore availability created major fishing pressure on this species by both recreational and commercial user groups. Commercial and recreational catches have declined significantly since the early 2000s and are at a fraction of their historic levels due to inshore habitat degradation and fishing pressure. Since the early 2000s, the southern New England/Mid-Atlantic stock of winter flounder has declined to perilously low levels.

Winter flounder historically has been a prime commercial target species because of its easy availability, consumer demand, and, in turn, consistent good market price. The principal commercial fishing gear for winter flounder is the otter trawl. Pound nets and fyke nets catch smaller amounts. The New York commercial catch includes winter flounder caught in New York waters (within 3 mi.) as well as fish caught outside New York waters.

Winter flounder has historically been important in New York commercial landings. As with many nearshore species in New York, winter flounder commercial catches were highest earlier in the twentieth century. New York landings peaked at 6.8 million pounds in 1938 then averaged around 3 million pounds through the 1940s. Landings were then around 1 million pounds per year until they began to drop significantly in the early 2000s as overall abundance and thus quotas declined. Landings are regulated by strict quota and remain very low as the resource rebuilds. Recent New York commercial landings have been less than 100,000 lb. Coast-wide (Maine through Delaware) landings peaked at 40 million pounds in 1981 and are currently only a fraction of that level.

The recreational harvest in New York was between 6 and 11 million pounds in the early to mid-1980s when winter flounder were among the top recreational species caught in New York. The harvest has declined since then also because of reduced abundance and resultant low quotas. The New York recreational harvest has been only less than 1,000 lb. in recent years. Coast-wide winter flounder recreational harvest peaked in 1982 and have also been very low in recent years.

ALL TACKLE WORLD RECORD: 7.0 lb., Fire Island, New York, 1986

NEW YORK RECORD: 7 lb. 3.5 oz., 1997

MANAGEMENT. Winter flounder is managed jointly by the ASMFC, the NEFMC, NMFS, and the NYSDEC. The ASMFC and NYSDEC manage the resource in state waters through the FMP for Inshore Stocks of Winter Flounder. The NEFMC manages winter flounder as part of the Northeast Multispecies FMP, which also includes a variety of other species including cod, windowpane, and yellowtail flounder. Both sets of FMPs are complementary. The fishery is managed as three separate stocks: southern New England/Mid-Atlantic (SNE/MA), which includes New York; Georges Bank; and Gulf of Maine. Unfortunately, owing to inshore spawning habitat loss and previous fishing effort, the SNE/MA stock is overfished, but because of strict harvest controls overfishing is not occurring.

The FMPs have implemented a rebuilding program for winter flounder stocks. This rebuilding program is based on overall quotas, minimum size, seasons, gear restrictions, trip limits, and individual allocations in the commercial fishery. The recreational fishery regulations include overall quotas, minimum size, seasons, and bag limits.

AMERICAN SOLES
Family Achiridae (*lenguados suelas*)

These are right-eyed flatfish and live in both marine and freshwater environments. This family has a total of 33 species, but only 5 occur along the Atlantic coast of the United States and only 1 is commonly found in our region. The mouth and eyes are small. The right-side pelvic fin is joined to the anal fin.

Hogchoker

Trinectes maculatus (suela tortilla)

Maine to the Gulf of Mexico and Venezuela, uncommon north of Cape Cod
Max. length = 8 in., common = 4 in.

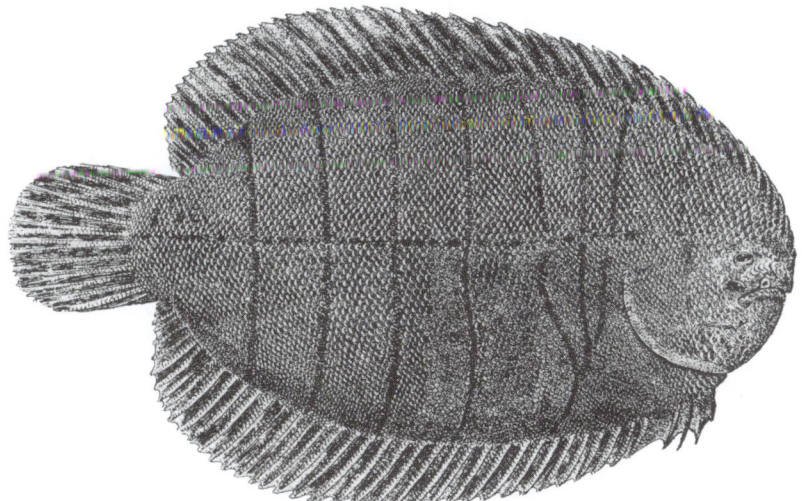

FIELD CHARACTERISTICS. More common in brackish water than other flatfishes. Eyed side is brown with blotches, spots, and about seven dark bars. The blind side often has dark blotches. The dorsal fin originates at the tip of the snout. No pectoral fins. The scales are rough (ctenoid) on both sides.

TRIGGERFISHES
Family Balistidae (*cochitos*)

The triggerfish family is commonly found in the tropical marine waters of the Indian, Pacific, and Atlantic Oceans. Although the 11 genera include about 40 species, only 1 species is routinely found in New York waters with 1 other species being relatively rare. The family is characterized by having a first dorsal fin consisting of only three spines with the first spine being conspicuously stout and large. When this first dorsal spine is erected, a bony knob on the base of the second dorsal spine fits into a groove on the back edge of the first spine, thus locking it in place. When the second spine is depressed, that acts as a "trigger" to release the locking mechanism. Other features shared by family members include a strongly compressed, rhomboidal body shape and a thick skin. Indeed, elsewhere triggerfish are called "leatherjackets."

Gray triggerfish

Balistes capriscus (pejepuerco blanco)

Western Atlantic coast Nova Scotia to Florida, Gulf of Mexico, Caribbean, Argentina, and eastern Atlantic
Max. length = 16 in.

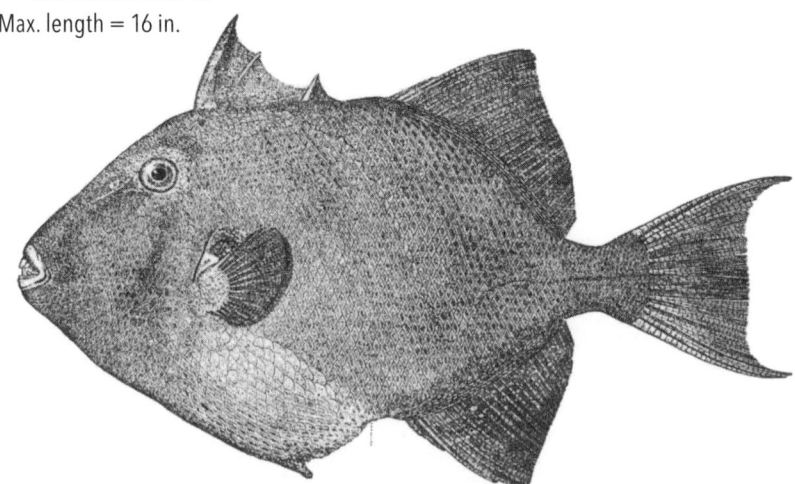

FIELD CHARACTERISTICS. Unlike the colorful queen triggerfish, a warm-water species rarely encountered during our summer, the gray triggerfish is predominantly gray with a white ventral area, although it has some blue spots and lines on the fins. Like the queen triggerfish, adults have long upper and lower caudal fin rays.

ECOLOGY/LIFE HISTORY. In general, coastal species found in the western Atlantic do not occur on the eastern side of the ocean. Somewhat uniquely, the gray triggerfish is found from the North Sea south to the Mediterranean and Angola.

Triggerfish belong to a group of fishes, the Tetradontiformes, which possess strong teeth. The gray triggerfish uses these teeth to seize and crush hard-shelled invertebrates such as crabs, mollusks, and sea urchins. In the tropics, this species also feeds on coral.

Triggerfish have a curious mode of swimming. Their second dorsal fin and anal fin are flexible and very similar in size. As the fish swims, those fins are flapped from side to side. To avoid predators, these fishes might swim into a crevice and lock their dorsal spine, making them difficult to be removed. Their large erect spine also serves simply to make the triggerfish too large to swallow easily. Not surprisingly, adult triggerfish have few predators.

FILEFISHES
Family Monacanthidae (*lijas*)

The serrated, slender first dorsal spiny ray is the most conspicuous feature in this large family of 32 genera and 102 species. Filefishes and triggerfishes are often confused, but the dorsal spine in filefishes is considerably weaker although longer. In both fishes, the spine can be locked in an upright position. The weak spine of the filefish may function as an anchor in juveniles or simply make the fish look more formidable to a potential predator. All the four species of filefishes collected within New York waters are juveniles or young adults and would be less than half their maximum lengths. These species, like many other young subtropical fishes, are recruited to New York waters through advection by the Gulf Stream.

Filefishes are generally seen as solitary, slow-moving browsers, nibbling on a variety of encrusting animals and plants. They have relatively small jaws but because their jaws are armed with strong incisor-like teeth, the larger filefish species can feed on small mollusks and coral polyps after breaking off pieces of coral. Corallivory is limited to a handful of fish families. The butterflyfish family has the greatest number of coral-feeding species. Other families possessing corallivores include puffers, triggerfishes, wrasses, parrotfishes, and damselfishes.

All filefishes have second dorsal fins and anal fins composed of numerous soft rays. Delicate waves of movement passing along these fins produce forces resulting in slow forward movement and stability. Filefishes have no pelvic fins.

In general, fishes with unusually shaped bodies, for example, eels, sand lances, triggerfishes, trunkfishes, and puffers, lack pelvic fins.

Orange filefish

Aluterus schoepfi (lija naranja)

Other common names: filefish, unicornfish
Nova Scotia, Florida, Gulf of Mexico, Caribbean, Brazil
Max. length = 2 ft.

FIELD CHARACTERISTICS. The orange filefish has orange dots on a variably patterned gray to brown background. This species is often found in habitats that contain erect objects, such as sea grasses, tube sponges, and horny corals. It is within these objects that this slow-swimming, secretive fish can hide from predators. The inconspicuous, laterally compressed shape, its camouflaged body color pattern, and its head-down posture serve to reduce detection by predators. Juvenile filefishes often are associated with *Sargassum* or under floating objects.

Planehead filefish

Stephanolepis hispidus (lija aspera)

Nova Scotia, Florida, Gulf of Mexico, Caribbean, Brazil
Max. length = 10 in.

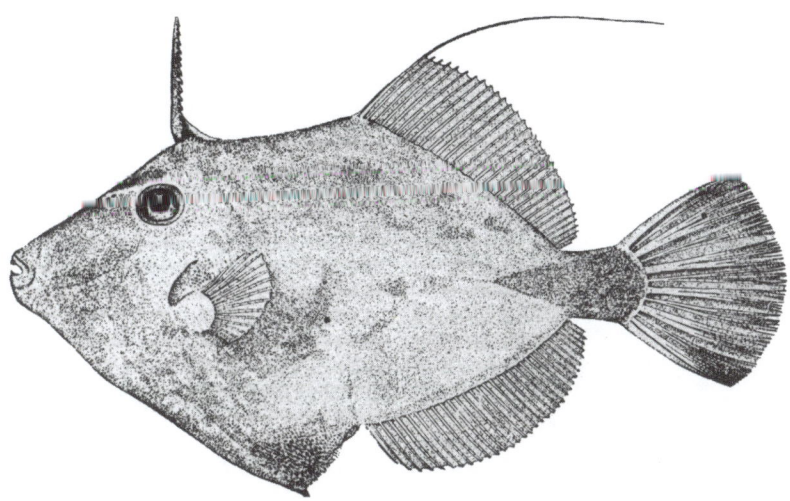

FIELD CHARACTERISTICS. Among the visiting filefishes, this species and the orange filefish appear to be the most commonly observed. In this species there is a fan-like pelvic flap extending from the end of the pelvic spine to the vent. Theoretically, this "pelvic dewlap" may serve to join a locked dorsal spine in anchoring the laterally compressed fish within a narrow crevice. The planehead also possess a prominent pelvic bone spine not seen in the other filefish discussed.

PUFFERS
Family Tetraodontidae (*botetes*)

This is a large, broadly distributed family with 112 species organized within 19 genera. They are predominantly marine but with some exceptions. This family owes its scientific name to four fused sets of beak-like teeth in its jaws. All members of the family are able to inflate their bodies when threatened. This ability involves the pumping of water from the mouth into a modified ventral portion of their stomach. This act is facilitated by a forward movement of the pectoral girdle that compresses the oral cavity, forcing water into an expandable portion of the stomach. Owing to the special organization of collagen and elastin fibers in the skin, puffer skin has a greater extensibility than most fishes, thus allowing a puffer to alter its shape.

Puffers have a relatively small mouth, but with their two large incisor-like teeth in the upper and lower jaws these fishes have a diverse diet that includes small crustaceans and mollusks.

Northern puffer

Sphoeroides maculatus

Other common names: blowfish, swellfish, balloonfish
Newfoundland to Florida
Max. length = 10 in.

FIELD CHARACTERISTICS. The northern puffer has vertically elongate bars of irregular length and width along the boundary between the dark dorsum and pale venter. The single dorsal and single anal fins are small and placed toward the caudal peduncle. The pectoral fin is immediately in front of the small gill opening, the latter being without a gill cover (operculum). There are no pelvic fins. This is probably a consequence of the fish's need to dramatically change its shape while inflated. The genus name *Sphoeroides* means spherical.

ECOLOGY/LIFE HISTORY. Puffers are generally small and slow moving and thus tempting prey. However, once inflated their increased size exceeds the gape of many potential predatory fish. Because the puffer can also pump air if taken out of the water, this ability might deter predatory birds as well. Puffers are predatory as well but their invertebrate prey is small.

Puffers are prized as food fishes because they are a tasty source of protein and are not often collected in great numbers. This species has been successfully cultured in laboratories on Long Island. In some parts of the world, the family is notorious for sequestering tetrodotoxin in their viscera, but the local puffers are not dangerously toxic although a different species, the bandtail puffer, is poisonous in Brazilian coastal waters.

FISHERIES. Although very desirable with a nearly bone-free edible portion, puffers do not support a substantial recreational or commercial fishery. Some of this is because their abundance numbers are not high enough to support a large commercial fishery and they do not put up a fight when caught on a hook. However, they are good eating, and there are niche markets for them in some local restaurants and fish markets. They are prized by locals who know their flavor.

The commercial fishery for puffers peaked in the 1940s at over two million pounds. This may have been because of successful year classes and a surge in their population. However, some of the reason for the high landings was the high demand for seafood during the World War II era when meat was being rationed.

Landings continued at around a half million pounds through the 1960s but have generally been less than 20,000 lb./yr. since 2000. Commercial landings are primarily by pound net and pots from inshore bays. Annual recreational catches have fluctuated wildly from less than 1,000 lb. to over 200,000 lb. since the early 1980s. However, these wide fluctuations are likely an aberration of the recreational survey.

MANAGEMENT. There is no FMP for puffers and thus no restrictions on either the commercial or the recreational fisheries. However, New York is considering the implementation of a minimum size in both the recreational and commercial fisheries.

Bandtail puffer

Sphoeroides spengleri (botete collarete)

Massachusetts to Florida, Gulf of Mexico, Caribbean, Brazil
Max. length = 7 in.

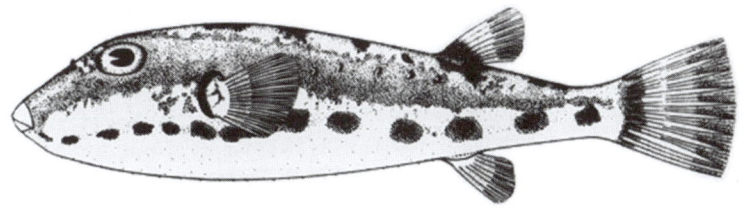

FIELD CHARACTERISTICS. This species is distinctly different from the northern puffer. The bandtail puffer has two dark bands on the tail and a row of dark blotches along the lower part of the body from chin to caudal fin base.

LITERATURE. Oliveira et al. 2003

PORCUPINEFISHES
Family Diodontidae (*peces erizo*)

This family of six genera and 19 species are warm-water fishes that can be found along the coasts of all the major oceans and many coral reefs.

Porcupinefishes are closely related to the puffer family and share its ability to inflate their bodies with water. An additional feature found in porcupinefishes is the array of immoveable sharp spines that cover the body and make the inflated fish even better protected against predators. Another difference is in the dentition. All members of this family have a fused tooth plate in the upper and lower jaws. These formidable teeth and jaws are used to crush corals and encrusting invertebrates.

Striped burrfish

Chilomycterus schoepfii (guanabana rayada)

Other common names: porcupinefish
Massachusetts to the Gulf of Mexico and Brazil
Max. length = 10 in.

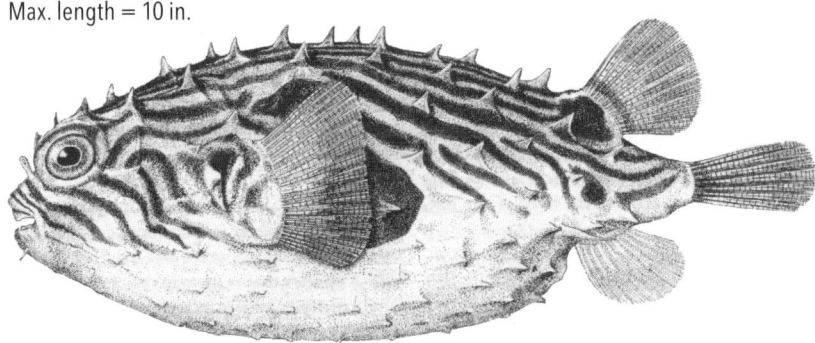

FIELD CHARACTERISTICS. These fish have an unusual pattern of wavy dark stripes on their dorsal surface and sides. The colors vary from yellowish green on the backside and pale on the belly. Further, there are dark eyespots (ocelli) arranged at different locations on the body. Eyespots are thought to confuse or intimidate predators.

The striped burrfish, like other diodontids and their puffer relatives, is oval in shape and has small gill openings and small fins. The body has reduced flexibility so the paddle-like pectoral fins, the single dorsal and anal fin, and the caudal fin are the primary means of locomotion and maneuvering. Curiously, a minor part of this fish's forward locomotion is due to propulsive

AND EXHALE

"*And exhale.*"

Cartoon by Michael Maslin, *New Yorker Magazine*, 2013

This fish seems to have been asked to inhale, hold its breath, and then exhale. Neither this member of the puffer fish family nor any fish, other than lungfish, have "lungs." Most fishes do have swimbladders, but, in general, that structure is not used for a respiratory purpose but to control a fish's buoyancy. So how does a puffer "inhale" and "exhale"?

In nature, this fish swims about normally until threatened. Then, in order to make themselves less available to a predator, they pump water into a modified part of their stomach. This is not something an ordinary fish can do. In the illustration, our fish "patient" is not under water. Can puffers inflate their bodies with air? Yes, they can.

forces that result when water is expelled through the small, round gill openings. Striped burrfish has no pelvic fins nor do any of the other fishes that are classified in the Order Tetraodontiformes.

LITERATURE. Wainright et al. 1995

SUNFISHES
Family Molidae (*molas*)

There are, in total, four species within this family. Three occur along the North American Atlantic coast but only one species occurs in local waters.

Ocean sunfish

Mola mola

Worldwide tropical and temperate waters, Newfoundland to the Gulf of Mexico and Argentina
Max. length = 11.7 ft., av. = 4.2 ft.

FIELD CHARACTERISTICS. Ocean sunfish are unmistakable surface-dwelling fish. They are the heaviest of all teleosts. An 8.3 ft. specimen weighed 1795 lb. Claims of longer specimens that are twice that weight have been made. Those sizes and their round or oval appearance is the primary reason why their name comes from the Latin for "millstone." The dorsal and anal fins are large and are responsible for the mola's locomotion. They have no true functional caudal fin. The posterior end of the body is reduced to a leathery flap that apparently originates from modified elements of the dorsal and anal fins rather than being a modified caudal fin. There are no pelvic fins. The mouth is small, and the upper and lower jaws each contain a single fused plate. The skin is very tough, nearly impenetrable.

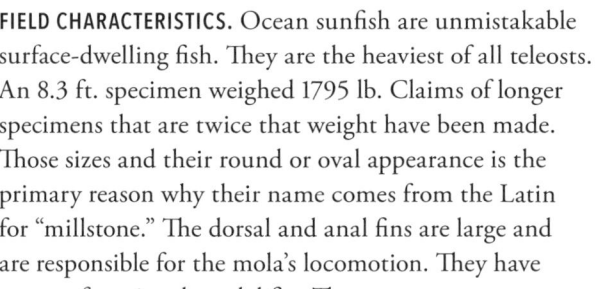

A WORD ABOUT SHELLFISHERIES

By design, this book is about finfishes and finfish fisheries. However, shell-fishes have historically been very important to New York's commercial fisheries. During the late 1800s and early 1900s, oysters (*Crassostrea virginica*) supported a very large fishery when 10 million to 15 million pounds were landed annually in New York. Peak landings of 20 million pounds occurred in 1904. Landings of just below 10 million pounds continued until the mid-1950s. Hard clams (*Mercenaria mercenaria*) supported a very large fishery during the 1960s and 1970s when approximately 75% of the East Coast hard clams came from New York. Landings ranged from 5 million to 8 million pounds annually during this period with peak landings of 9 million pounds in 1976. The surf clam (*Spisula solidissima*) fishery became important in the 1950s when landings were 3 million to 5 million pounds annually. There was a resurgence in the fishery in the 1980s through the early 2000s when landings were 5 million to 10 million pounds annually. Peak landings of 14.5 million pounds occurred in 1993. The lobster (*Homarus americanus*) fishery greatly expanded during the 1980s and 1990s when landings went from 1 million to over 8 million pounds. Peak landings of 9.4 million pounds occurred in 1996 and have dropped off significantly since. Bay scallops (*Aequipecten irradians*) for many decades were an important fishery from the 1950s through the mid-1980s. Peak landings of just under 1 million pounds occurred in 1962.

All the above shellfisheries have experienced significant declines since their heydays. However, since the 1980s, longfin squid (*Doryteuthis (Amerigo) pealeii*) has become an important New York fishery. Since that time landings have generally been 4 million to 6 million pounds per year with peak landings of 13.2 million pounds in 2000. Longfin squid is currently within the top five commercially landed species in New York and is a staple of the trawl fishery.

Fisheries Management

The Native peoples who lived along New York's coastal waters may not have intended to manage their fisheries, but they did develop skills to harvest fishes in such a way that appears to have been sustainable. Much of what we know about the precolonial Indigenous people of Long Island was reviewed by John A. Strong (1997) in *The Algonquian Peoples of Long Island from Earliest Times to 1700*.

Some of the earliest evidence comes from small settlements of Native people on the western bank of Wading River and on Three Mile Harbor in East Hampton. During the earliest time (Archaic Period, 8000–3000 BP) when Long Island was first populated, it appears that the Indigenous people fished for sturgeon, striped bass, flounder, shad, and bluefish, among other fishes, by means of a variety of techniques that employed canoes, deer bone harpoons, and fishing nets. Smaller fishes were collected by cast nets thrown by hand. Grooved stones may have been used as sinkers. Fish weirs were also used. These weirs were made from wooden stakes and brush woven in between the stakes. Construction workers in an area adjacent to the Charles River estuary in Boston uncovered thousands of 4000-year-old wooden stakes that were part of large fish weirs used by Late Archaic Period Native people. The Boston Museum of Science Library displayed what that fish weir looked like, which is illustrated in C. Keith Wilbur's *The New England Indians*. Other illustrations in that volume include the use of branches to fence off narrow streams, thereby blocking fish, which are then speared; employment of stone walls to funnel downstream-moving fish to an awaiting collector; and the use of long nets made of hemp as gill nets. In creeks, after fishes came in on the high tide, the exit was blocked during the ebb, permitting easy harvesting of dense schools at low tide. It was probably very common for villages to be established along the banks of freshwater streams where they flowed into tidal waters. Late Archaic Period archaeological sites have been found on Long Island near Orient, James-

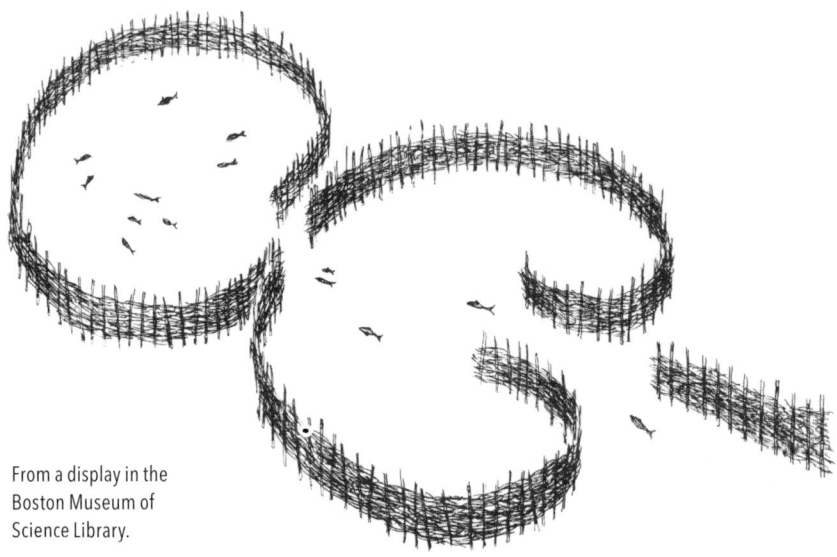

From a display in the
Boston Museum of
Science Library.

port, Wading River, and Stony Brook on the North Shore and Sugar Loaf near
Shinnecock Bay on the South Shore (Braun and Braun 1994).

In the Early and Middle Woodland Periods (3000–1000 BP) finfish and
shellfish were the primary sources of year-round protein. During the fall and
winter, bluefish, striped bass, blackfish, flounder, and sturgeon were caught.
Winter flounder in particular must have been easy prey, considering that they
spawn in shallow water in winter months. In the spring, dense schools of anadromous fishes such as alewives and shad migrate up into freshwaters streams
where they could be easily collected. Most of the species of food fishes that were
available in these early times are still in New York waters but are not as abundant as they once were.

In the Late Woodland Period (1000–600 BP) Narragansett people used
fishes, probably alewives, shad, or menhaden, as fertilizer. Those fishes were
buried just below the corn seed. Native people of Cape Cod used some other
species (e.g., silver hake and cod). It is likely that the Algonquians on Long Island used fish as fertilizer, but at present there is no direct evidence for this. It is
known that the Native peoples did use fish more imaginatively by adding fish,
beans, and corn to make chowders.

During the mid-seventeenth century, the Native people ceded some of the
first lands to the English settlers. Wyandanch, one of the most influential sa-
chems, was fully aware of the importance of fishes and fishing. He made sure
that the Native peoples of eastern Long Island, the Shinnecocks, Mantauks,
Manhassets, and Corchaugs, reserved the right to fish on those lands leased to
the settlers.

MODERN FISHERIES MANAGEMENT

Fisheries management—it's a term we have all heard or one with which we have contended. But what is fisheries management and what does it mean? We are all familiar with some of the implementation actions of fisheries management, such as quotas, mesh size, bag limits, minimum fish sizes, and closed seasons. But how were these actions derived and what do they mean to the resource— the fish themselves?

The object of fisheries management is to sustain the resource—to ensure that there are fish enough for us to catch and to prevent depleting the resource. Additionally, we know that many fish species go through cycles of high and low abundance. Fisheries management strives to spread out the catch of the high part of the cycle in order to temper the boom-and-bust fishing cycles that occur in cyclic fisheries. Fisheries resources along the Middle-Atlantic and New England coasts experience significant fishing pressure by both commercial and recreational fishermen. The fishing pressure by both commercial and recreational fishermen is such that for most of our fishery resources, fishing mortality, or the total amount of fish killed by fishing activities, has to be limited.

The United States manages its fisheries in the Exclusive Economic Zone (EEZ; 3 to 200 mi. offshore) through the National Marine Fisheries Service (NMFS). The enabling legislation is the Magnuson–Stevens Act (as amended), originally signed into law in 1976. The Magnuson–Stevens Act established eight regional fisheries management councils to advise NMFS on fisheries management and to develop fishery management plans (FMPs) for the conservation and utilization of our nation's marine resources. Most fishes caught by New York fishermen in the EEZ come under the jurisdiction of the Mid-Atlantic Fishery Management Council (MAFMC), which includes the states from New York to North Carolina. These include bluefish, scup, black sea bass, summer flounder, and others. More northern species such as cod, pollock, winter flounder, and silver hake fall under the jurisdiction of the New England Fishery Management Council (NEFMC), which includes the states from Maine to Connecticut.

Fisheries within 3 mi. of the coast are managed by the individual states. However, the 15 Atlantic coastal states from Maine to Florida have come together to form the Atlantic States Marine Fisheries Commission (ASMFC). The ASMFC develops FMPs, which the member states then implement in their respective states. The ASMFC also works with the MAFMC (and NEFMC) in developing FMPs for species that overlap jurisdictions of the two management bodies.

Although the MAFMC, NEFMC, and the ASMFC are separate entities, they work together with the states on the development of fishery management plans, including stock assessments, quota setting, and other management measures. In fact, many managers are members of both the council and the commission.

In New York State, fisheries regulations are developed to implement council and commission FMPs within state waters. This often is a complicated process but starts with the New York Legislature. The legislature enacts laws for each regulated species by three different methods:

1. The regulation specifics (e.g., minimum size, seasons) are set by the law and can be changed only through legislative action.
2. The legislature, by law, gives the New York State Department of Environmental Conservation (NYSDEC) regulatory authority to set the specific regulations usually based on a specified procedure, including public hearings. Thus, the NYSDEC can proceed and set and/or change regulations.
3. A combination of the preceding two: the legislature will set certain regulations in law while giving the NYSDEC the authority to also set some of the regulations.

Some species are also regulated by international commissions. These primarily include pelagic species that cross international boundaries, such as tunas and billfishes.

So how do fishery managers go about the business of managing fisheries and what is a fishery? A fishery can be defined as a species or a stock of fish (or a group of stocks) and all the activities involved in catching those fish, including recreational fishing, commercial fishing, boats and vessels, bait and tackle, gear manufacture, processing, distribution, and sales. A fish stock is a management unit based on a grouping of a fish species that can be based on genetics, geographic distribution, spawning, or migration. Let us graphically represent the total amount of fish in a particular stock of fish as a box. The two things that add to the stock of fish and make the box bigger are births of more fish and growth of the fish in the population. The two things that decrease the stock of fish and make the box smaller are natural mortality and fishing mortality. Natural mortality includes death due to old age, predation, disease, or pollution; the rate of natural mortality is denoted by fisheries managers as M. For most fish stocks natural mortality is highest from the egg to larval stages. As fish pass their first year, these natural causes of death usually decline, and in many cases fishing becomes the dominant source of mortality. Fishing mortality includes fish harvested by both commercial and recreational fishermen as well as discard mortality—fish that die after being released alive or those released dead. Fish that are caught and brought ashore are referred to as harvest or landings. When discard mortality is added in, the total is designated as catch. Catch determines fishing mortality, and the rate of fishing mortality caused by recreational and commercial fisheries is denoted by fisheries managers as F. Total mortality, Z, is the sum of F and M.

An FMP is exactly what its name suggests: a plan to successfully manage a fishery for continued sustainability. The goal of an FMP is to protect the resource. The associated objectives are to prevent overfishing, rebuild the resource if it is overfished, provide the greatest good to resource users by achieving optimum yield from the resource, and maintain and fairly allocate fisheries through cooperative regulatory planning. Fishery management plans contain information on the biology of the stock and on the fishery. The FMP identifies problems and recommends conservation and management measures to address problems and prevent overfishing or rebuild the resource and meet management objectives. The FMPs are not static and are modified and updated as needed, usually if a stock assessment indicates a change in stock status. An FMP can be modified by amendments, addendums, and frameworks. The process is transparent and takes place in a public forum; fishermen are encouraged to participate in the process.

The study of a population of fish as a living unit is called population dynamics. The term population dynamics refers to the rate of reproduction, growth, and death of a population of fish. Age structure is the percentage of each age group in the total population. Over time, the size and age composition of a fish population is always changing. Data on reproduction, growth, death (F and M), and age structure is what fishery scientists use to develop a mathematical model that represents a fish population. Population dynamics uses mathematical models to explain and predict changes in populations over time. This prediction is accomplished through a fisheries stock assessment process. A stock assessment uses a variety of biological, scientific, and fishery-dependent and fishery-independent information and data to determine the status of the stock relative to overfishing or being overfished (see below). Stock assessments are updated on a regular basis for each species.

The stock assessment process helps to inform the FMP about important parameters of the stock status: births (strength of new year classes), growth, the size of the stock of fish, and removals by fishing and natural mortality. Fisheries managers then use the information from the stock assessment to compare it with limits, targets, and thresholds established in the FMP to meet its objectives. The stock assessment also provides information to fisheries managers to evaluate how stocks have been affected by recent management actions and to forecast future conditions under various management actions.

Two important parameters that come out of the stock assessment are abundance and exploitation. **Abundance** can be in numbers of fish but is usually defined as the biomass or total weight of the stock of fish. Spawning stock biomass is also often used and denotes the weight of mature fish that can spawn or, more often, the total weight of mature females in the population. The FMP develops a standard, or reference point, against which to compare the biomass level to determine the status of the stock. If the biomass is below the threshold reference point (the minimum biomass needed to maintain the population at a sustainable level) then biomass is too low and the stock is considered over-fished. New conservation and management measures must then be developed to rebuild the resource. If the biomass is above the threshold, then it is not overfished. Many FMPs also develop a target biomass level that is greater than the threshold. Managers try to build the biomass to the target level to provide a buffer for the stock.

Exploitation refers to the mortality caused by fishing and is denoted as F and is expressed as a rate of exploitation. It includes both fish that are landed as well as the dead discards and those that will die after being discarded. As with biomass, the FMP establishes a standard or a reference point against which to compare F to determine stock status. If F is above the threshold reference point, then the exploitation rate is too high and overfishing is occurring. New conservation and management measures need to be developed to reduce fishing effort. If F is below the threshold, then overfishing is not occurring. As with biomass, many FMPs also develop a target F that is lower than the threshold. Managers try to reduce fishing effort to the target level to provide a buffer for the stock. Note that fishing effort can be reduced relatively quickly by reducing quotas, bag and trip limits, and seasons. However, it can take years to rebuild biomass. So, it is possible that a stock can still be overfished (rebuilding) and not experiencing overfishing.

Mortality rates are critical for determining the abundance of fish populations and the effect of harvesting strategies on yield and spawning potential from a stock of fish. The rate at which a stock is harvested is usually estimated by calculating the abundance of a year class over successive years to determine how fast it is declining. Fishing mortality can be changed through indirect methods, such as regulating mesh size to make fish of certain ages less vulnerable to the gear. Direct control measures, such as catch quotas or effort limits, determine the rate of fishing mortality on the vulnerable sizes.

We need to keep in mind that overfishing occurs as the result of the cumulative impact of many commercial and recreational fishermen. Even if we think that our personal catch is small or insignificant, we need to remember that the cumulative impact of many small catches, as well as discards, can be significant. By monitoring and regulating fishing mortality rates, managers may provide acceptable protection against overfishing for a particular stock.

Each one of us in our own way, be it commercial or recreational, adds to the cumulative effect of harvesting fish. We therefore each in our own way have the ability to help contribute to fisheries management and to help insure a continued abundance of fish for the future. Fish stocks are not limitless. We can all share in the benefits of harvest as well as bear the responsibility for future abundance. Commercial and recreational fishermen need to work together toward that end.

Literature: Kilduff et al. 2009

Who Manages Which Fishery?

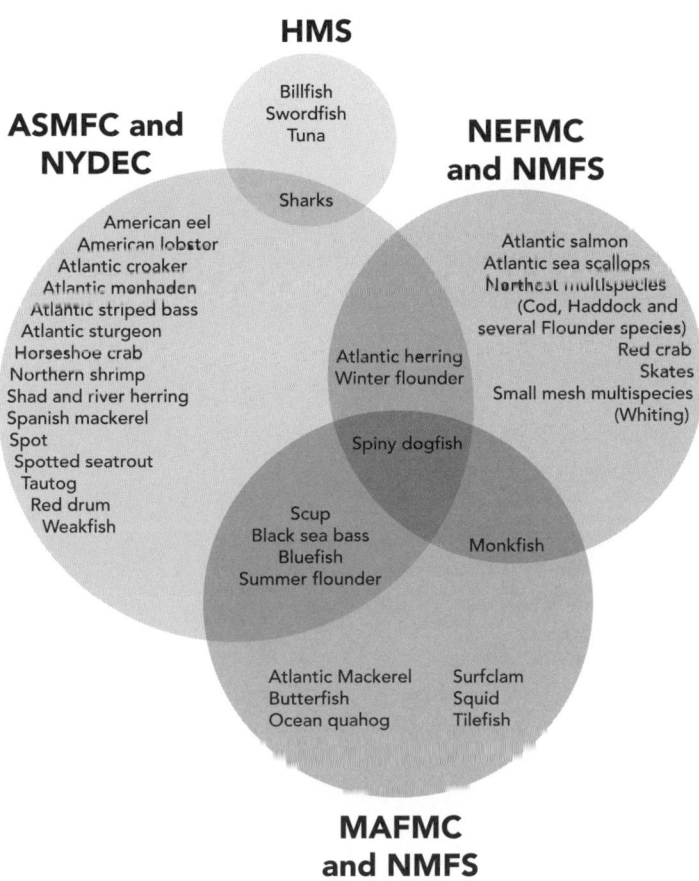

HMS

Billfish
Swordfish
Tuna

Sharks

**ASMFC and
NYDEC**

American eel
American lobster
Atlantic croaker
Atlantic menhaden
Atlantic striped bass
Atlantic sturgeon
Horseshoe crab
Northern shrimp
Shad and river herring
Spanish mackerel
Spot
Spotted seatrout
Tautog
Red drum
Weakfish

**NEFMC
and NMFS**

Atlantic salmon
Atlantic sea scallops
Northeast multispecies
(Cod, Haddock and
several Flounder species)
Red crab
Skates
Small mesh multispecies
(Whiting)

Atlantic herring
Winter flounder

Spiny dogfish

Scup
Black sea bass
Bluefish
Summer flounder

Monkfish

Atlantic Mackerel   Surfclam
Butterfish   Squid
Ocean quahog   Tilefish

**MAFMC
and NMFS**

Adapted from NMFS 2016.

CONSERVATION GEAR TECHNOLOGY

Many people involved in fisheries management and fisheries research, including author ECH, are also very involved in conservation gear technology (CGT). Conservation gear technology works to develop and improve fishing gear technology that helps to reduce bycatch and discards. Bycatch are fish that are caught in a fishery but are not retained for reasons such as they are outside the legal size limit; there is no market demand for that species or that size of fish; the fisherman has already caught their trip limit or bag limit; the quality of the fish is poor; regulations do not allow fisherman to retain the fish because of closed seasons, closed quota, use of a net mesh size that is legal for some species but not for other species; or the species is not allowed to be landed (including threatened or endangered species), as well as other reasons. Bycatch is typically discarded, some dead, some alive. Even for those fish discarded alive, some percentage will die.

Conservation gear technology works to develop more efficient fishing gear that retains the target species while at the same time allows bycatch to escape the gear while the gear is still fishing or minimizes injury to bycatch to reduce discard mortality.

Conservation gear technology includes

- determination of net mesh sizes that retain legal size fish but allow undersize fish to escape through the meshes;
- determination of design and installation of escape panels in lobster and fish pot gear that allows small fish and lobsters to escape and that will also corrode and fall off over time so the pot does not continue to fish if it becomes a lost or "ghost" pot;
- modifications in trawl gear design and construction that uses fish behavior and various size meshes and openings to allow bycatch to escape while the net is fishing;
- hook design that results in a fish being hooked in the lip or jaw rather than gut-hooked to reduce the chance of death after discard; and
- fish handling techniques to reduce injury.

As these CGT approaches are developed, many are incorporated into fishing regulations. Mesh-size regulations for trawls, which are often different for different species, are required by fishery management plans. Large-mesh panels strategically placed in small-mesh trawls allow nontarget species to escape. Raising the bottom of a trawl slightly off the seabed allows some species to avoid capture. Escape panels for pot gear are required in many areas. Circle hooks are now required in the recreational striped bass fishery when fishing with bait.

Author ECH and his fisheries team coworkers at the Cornell Cooperative Extension Marine Program have been involved in CGT research with the fishing industry for many years. Results of these various projects can be found at https://ccesuffolk.org/marine/fisheries. We have helped to develop many of the CGTs mentioned above. We have also developed a fishing fleet avoidance program whereby fishermen report from the fishing grounds if they encounter high concentrations of certain species that should be avoided to reduce bycatch. That information is then provided back to the fishing fleet so other fishermen can avoid those specific areas and thus reduce bycatch.

APPENDIX

Marine Fishes of New York
and Their Relative Occurrence

In general, there are four categories of fishes. These categories and their estimated numbers are as follows. Jawless fishes (70 lampreys, 35 hagfishes); cartilaginous fishes (970 sharks, skates, and rays), bony lobe-finned fishes (2 coelacanths, 6 lungfishes), and bony ray-finned fishes (2 paddlefishes, 25 sturgeons, 16 bichirs, 7 gars, 1 bowfin, and 26,840 teleost species). The marine fishes most commonly encountered in New York marine waters are teleosts. Worldwide, except for teleosts, the number of identified species will likely not change significantly. However, every year new teleosts are being discovered and named as the result of explorations in the deep sea, tropical rivers and lakes, and isolated coral reefs.

The marine waters of New York State include the tidal reaches of the Hudson River, New York Harbor, and to the greatest extent the marine habitats within and surrounding Long Island. Long Island is 118 mi. long and has an estimated 1600 mi. of shoreline defined by many large and diverse harbors, bays, and coastal areas. The distribution of the marine fish species within these water bodies is uneven and complex. Any attempt to rank their relative abundance will be imperfect. For instance, those species categorized as "rare" can be reliably found in certain locations at particular times of the year. Many more rare species are not predictably encountered and are listed secondarily. The subjective judgments made here are based on the authors' personal experiences and results from collections documented in the scientific literature.

AGNATHA (JAWLESS FISHES)

Petromyzontidae
Petomyzon marinus sea lamprey (common)

CHONDRICHTHYES (CARTILAGINOUS FISHES)

Odontaspididae
Carcharias taurus sand tiger (common)
Alopiidae
Alopias vulpinus common thresher shark (not uncommon)
Cetorhinidae
Cetorhinus maximus basking shark (not uncommon)
Lamnidae
Carcharodon carcharias white shark (not uncommon)
Isurus oxyrinchus shortfin mako (common)
Lamna nasus porbeagle (occasional)
Triakidae
Mustelus canis smooth dogfish (abundant)
Carcharinidae
Carcharinus obscurus dusky shark (common)
Carcharinus plumbeus sandbar shark (common)
Galeocerdo cuvier tiger shark (not uncommon)
Prionace glauca blue shark (common)
Rhizoprionodon terraenovae Atlantic sharpnose shark (rare)
Sphyrnidae
Sphyrna zygaena smooth hammerhead (not uncommon)
Squalidae
Squalus acanthias spiny dogfish (abundant)
Torpedinidae
Torpedo nobiliana Atlantic torpedo (common)
Rajidae
Dipturus laevis barndoor skate (not uncommon)
Leucoraja erinacea little skate common)
Leucoraja ocellata winter skate (common)
Raja eglanteria clearnose skate (common)
Dasyatidae
Dasyatis centroura roughtail stingray (common)
Myliobatidae
Rhinoptera bonasus cownose ray (erratic)

ACTINOPTERYGII (RAY-FINNED BONY FISHES)

Acipenseridae
  Acipenser brevirostrum shortnose sturgeon (rare)
  Acipenser oxyrinchus Atlantic sturgeon (formerly common)
Anguillidae
  Anguilla rostrata American eel (abundant)
Congridae
  Conger oceanicus conger eel (common)
Engraulidae
  Anchoa hepsetus striped anchovy (erratic)
  Anchoa mitchilli bay anchovy (abundant)
Clupeidae
  Alosa aestivalis blueback herring (common)
  Alosa mediocris hickory shad (common)
  Alosa pseudoharengus alewife (common)
  Alosa sapidissima American shad (common)
  Brevoortia tyrannus Atlantic menhaden (abundant)
  Clupea harengus Atlantic herring (abundant)
  Dorosoma cepedianum gizzard shad (rare)
  Etrumeus teres round herring (common but uncommon in bays)
Osmeridae
  Osmerus mordax rainbow smelt (erratic)
Synodontidae
  Synodus foetens inshore lizardfish (sometimes common in summer)
Merluccidae
  Merluccius bilinearis silver hake (common)
Phycidae
  Enchelyopus cimbrius fourbeard rockling (common in LI Sound)
  Urophycis chuss red hake (abundant)
  Urophycis regia spotted hake (not uncommon)
  Urophycis tenuis white hake (common)
Gadidae
  Gadus morhua Atlantic cod (common)
  Microgadus tomcod Atlantic tomcod (locally common)
  Melanogrammus aeglefinus haddock (uncommon)
  Pollachius virens pollock (common)
Ophidiidae
  Ophidion marginatum striped cusk-eel (common)

Batrachoididae
　　Opsanus tau oyster toadfish (abundant)
Lophiidae
　　Lophius americanus goosefish (common)
Mugilidae
　　Mugil cephalus striped mullet (common)
　　Mugil curema white mullet (common)
Atherinopsidae
　　Membras martinica rough silverside (common)
　　Menidia beryllina inland silverside (common)
　　Menidia menidia Atlantic silverside (abundant)
Hemiramphidae
　　Hyporhamphus meeki false silverstripe halfbeak (rare)
Belonidae
　　Strongylura marina Atlantic needlefish (common in summer)
Scomberesocidae
　　Scomberesox saurus Atlantic saury (uncommon)
Fundulidae
　　Fundulus heteroclitus mummichog (abundant)
　　Fundulus luciae spotfin killifish (locally common)
　　Fundulus majalis striped killifish (abundant)
　　Lucania parva rainwater killifish (common)
Cyprinodontidae
　　Cyprinodon variegatus sheepshead minnow (abundant)
Zeidae
　　Zenopsis conchifera buckler dory (uncommon)
Gasterosteidae
　　Apeltes quadracus fourspine stickleback (abundant)
　　Gasterosteus aculeatus threespine stickleback (abundant)
　　Gasterosteus wheatlandi blackspotted stickleback (uncommon)
　　Pungitius pungitius ninespine stickleback (locally common)
Syngnathidae
　　Hippocampus erectus lined seahorse (common)
　　Syngnathus fuscus northern pipefish (abundant)
Fistulariidae
　　Fistularia tabacaria bluespotted cornetfish (uncommon)
Scorpaenidae
　　Pterois volitans red lionfish (rare)

Triglidae
  Prionotus carolinus northern searobin (abundant)
  Prionotus evolans striped searobin (abundant)

Cottidae
  Myoxocephalus aenaeus grubby (common)
  Myoxocephalus octodecemspinosus longhorn sculpin (common)
  Myoxocephalus scorpius shorthorn sculpin (rare)

Hemitripteridae
  Hemitripterus americanus sea raven (common)

Cyclopteridae
  Cyclopterus lumpus lumpfish (not uncommon)

Moronidae
  Morone americana white perch (common)
  Morone saxatilis striped bass (common)

Serranidae
  Centropristis striata black sea bass (common)

Malacanthidae
  Lopholatilus chamaeleonticeps tilefish (common)

Pomatomidae
  Pomatomus saltatrix bluefish (abundant)

Carangidae
  Alectis ciliaris African pompano (uncommon)
  Caranx chrysos blue runner (sometimes common)
  Caranx hippos crevalle jack (juvenile common in summer)
  Decapterus macarellus mackerel scad (sometimes common)
  Decapterus punctatus round scad (sometimes common)
  Oligoplites saurus leatherjack (uncommon)
  Selar crumenophthalmus bigeye scad (not uncommon)
  Selene setapinnis Atlantic moonfish (not uncommon)
  Selene vomer lookdown (juvenile not uncommon)
  Seriola dumerili greater amberjack (uncommon)
  Seriola zonata banded rudderfish (common)
  Trachinotus carolinus Florida pompano (not uncommon)
  Trachinotus falcatus permit (juvenile common in summer)
  Trachurus lathami rough scad (not uncommon)

Rachycentridae
  Rachycentron canadum cobia (uncommon)

Coryphaenidae
  Coryphaena hippurus dolphinfish (rare)

Echeneidae
  Echeneis naucrates sharksucker (not uncommon)
Lutjanidae
  Lutjanus griseus gray snapper (juveniles not uncommon in summer)
Sparidae
  Archosargus probatocephalus sheepshead (uncommon)
  Lagodon rhomboides pinfish (uncommon)
  Stenotomus chrysops scup (abundant)
Sciaenidae
  Bairdiella chrysoura silver perch (sometimes common)
  Cynoscion regalis weakfish (common)
  Leiostomus xanthurus spot (erratic)
  Menticirrhus saxatilis northern kingfish (common)
  Micropogonias undulatus Atlantic croaker (uncommon)
Chaetodontidae
  Chaetodon capistratus foureye butterflyfish (rare)
  Chaetodon ocellatus spotfin butterflyfish (juvenile not uncommon in summer)
  Chaetodon striatus banded butterflyfish (rare)
Labridae
  Tautoga onitis tautog (abundant)
  Tautogolabrus adspersus cunner (abundant)
Zoarcidae
  Zoarces americanus ocean pout (common)
Pholidae
  Pholis gunnellus rock gunnel (common)
Ammodytidae
  Ammodytes americanus American sand lance (abundant)
Uranoscopidae
  Astroscopus guttatus northern stargazer (uncommon)
Blenniidae
  Hysoblennius hentz feather blenny (uncommon)
Gobiidae
  Gobiosoma bosc naked goby (common)
  Gobiosoma ginsburgi seaboard goby (locally common)
Acanthuridae
  Acanthurus chirurgus doctorfish (rare)
  Acanthurus coeruleus blue tang (rare)
  Acanthurus tractus ocean surgeonfish (uncommon)

Sphyraenidae
  Sphyraena borealis sennet (common in summer)
Scombridae
  Euthynnus alletteratus little tuny (irregular)
  Katsuwonas pelamis skipjack tuna (sometimes common)
  Sarda sarda Atlantic bonito (common)
  Scomber colias Atlantic chub mackerel (sometimes common)
  Scomber scombrus Atlantic mackerel (abundant seasonally)
  Scomberomorus maculatus Spanish mackerel (sometimes common)
  Thunnus alalunga albacore (irregular)
  Thunnus albacares yellowfin tuna (irregular)
  Thunnus obesus bigeye tuna (irregular)
  Thunnus thynnus bluefin tuna (common)
Xiphiidae
  Xiphias gladius swordfish (not uncommon)
Istiophoridae
  Kajikia albida white marlin (not uncommon)
  Makaira nigricans blue marlin (uncommon)
Stromateidae
  Peprilus triacanthus butterfish (not uncommon)
Scophthalmidae
  Scophthalmus aquosus windowpane (abundant)
Paralichthyidae
  Etropus microstomus smallmouth flounder (common)
  Paralichthys dentatus summer flounder (abundant)
  Paralichthys oblongus fourspot flounder (common)
Pleuronectidae
  Limanda ferruginea yellowtail flounder (common)
  Pseudopleuronectes americanus winter flounder (abundant)
Achiridae
  Trinectes maculatus hogchoker (common)
Balistidae
  Balistes capriscus gray triggerfish (common)
Monacanthidae
  Aluterus schoepfi orange filefish (uncommon)
  Stephanolepis hispidus planehead filefish (uncommon)
Tetraodontidae
  Sphoeroides maculatus northern puffer (common)
  Sphoeroides spengleri bandtail puffer (uncommon)

Diodontidae
Chilomycterus schoepfii striped burrfish ((not uncommon)
Molidao
Mola mola ocean sunfish (not uncommon in summer)

OTHER RARE SPECIES MORE COMMON IN WARMER WATERS

Sphyrnidae
Sphyrna lewini scalloped hammerhead
Sphyrna tiburo bonnethead
Gymnuridae
Gymnura micrura smooth butterfly ray
Myliobatidae
Myliobatis freminvillei bullnose ray
Mobulidae
Manta birostris giant manta
Elopidae
Elops saurus ladyfish
Megalopidae
Megalops atlanticus tarpon
Clupeidae
Opisthonema oglinum Atlantic thread herring
Salmonidae
Salmo salar Atlantic salmon
Oncorhynchus kisutch coho salmon
Synodontidae
Trachinocephalus myops snakefish
Antennariidae
Histrio histrio sargassumfish
Antennarius striatus striated frogfish
Exocoetidae
Cheilopogon melanurus Atlantic flyingfish
Belonidae
Ablennes hians flat needlefish
Tylosurus crocodilus houndfish
Holocentridae
Holocentrus adscensionis squirrelfish
Aulostomidae
Aulostomus maculatus Atlantic trumpetfish

Dactylopteridae
   Dactylopterus volitans flying gurnard
Scorpaenidae
   Scorpaena plumieri spotted scorpionfish
Epinephelidae
   Cephalopholis fulva coney
   Epinephelus adscensionis rock hind
   Epinephelus guttatus red hind
   Epinephelus itajara Atlantic goliath grouper
   Epinephelus morio red grouper
   Hyporthodus nigritus Warsaw grouper
   Hyporthodus niveatus snowy grouper
   Mycteroperca bonaci black grouper
   Mycteroperca interstitialis yellowmouth grouper
   Mycteroperca microlepis gag
   Mycteroperca phenax scamp
   Paranthias furcifer Atlantic creolefish
Serranidae
   Baldwinella aureorubens streamer bass
   Diplectum formosum sand perch
Apogonidae
   Apogon pseudomaculatus twospot cardinalfish
Priacanthidae
   Heteropriacanthus cruentatus glasseye snapper
   Priacanthus arenatus bigeye
   Pristigenys alta short bigeye
Carangidae
   Caranx bartholomaei yellow jack
   Caranx latus horse-eye jack
   Chloroscombrus chrysurus Atlantic bumper
   Elegatis bipinnulata rainbow runner
   Naucrates ductor pilotfish
   Seriola fasciata lesser amberjack
   Trachinotus goodei palometa
Lutjanidae
   Lutjanus analis mutton snapper
   Lutjanus apodus schoolmaster
   Lutjanus cyanopterus cubera snapper
   Lutjanus mahogoni mahogony snapper

Lobotidae
  Lobotes surinamensis Atlantic tripletail
Gerreidae
  Eucinostomus gula silver jenny
Polynemidae
  Polydactylus octonemus Atlantic threadfin
Haemulidae
  Haemulon flavolineatum French grunt
Sciaenidae
  Pogonias cromis black drum
  Sciaenops ocellatus red drum
Mullidae
  Mullus auratus red goatfishs
  Pseudupeneus maculatus spotted goatfish
  Upeneus parvus dwarf goatfish
Chaetodontidae
  Chaetodon sedentarius reef butterflyfish
Pomacanthidae
  Holacanthus bermudensis blue angelfish
  Pomacanthus arcuatus gray angelfish
  Pomacanthus paru French angelfish
Pomacentridae
  Abudefduf saxatilis sergeant major
  Stegastes leucostictus beaugregory
  Stegastes partitus bicolor damselfish
Labridae
  Halichoeres bivittatus slippery dick
  Sparisoma rubripinne yellowtail parrotfish
  Xyrichtys novacula pearly razorfish
Gobiesocidae
  Gobiesox strumosus skilletfish
Ephippidae
  Chaetodipterus faber Atlantic spadefish
Sphyraenidae
  Sphyraena guachancho guaguanche
Trichiuridae
  Trichiurus lepturus Atlantic cutlassfish

Scombridae
  Scomber colias Atlantic chub mackerel
  Scomberomorus cavalla king mackerel
  Scomberomorus regalis cero
Centrolophidae
  Hyperoglyphe perciformis barrelfish
Stromateidae
  Peprilus paru harvestfish
Bothidae
  Bothus robinsi twopost flounder
Balistidae
  Balistes vetula queen triggerfish
Monacanthidae
  Aluterus heudelotii dotterel filefish
  Aluterus scriptus scrawled filefish
  Monacanthus ciliatus fringed filefish
Ostraciidae
  Acanthostracion quadricornis scrawled cowfish
  Lactophrys bicaudalis spotted trunkfish
  Lactophrys trigonus trunkfish
  Lactophrys triqueter smooth trunkfish
Tetradontidae
  Lagocephalus laevigatus smooth puffer

References

LITERATURE CITED

The fish names placed in parentheses after some of the citations identify the profile within which the reference was used if not otherwise obviously stated within the title of the reference.

Within many of the species profiles, references from the primary literature were used as a source for particular factual statements. That literature is noted, in abbreviated form, at the end of the profile. The full citations are in the following list. We did not attempt to specifically cite the many different sources used to determine standard information such as size, distribution, ecology, life history, or fisheries details.

Aalbers, S.A., D. Bernal, and C.A. Sepulveda. 2010. The functional role of the caudal fin in the feeding ecology of the common thresher *Alopias vulpinus*. J. Fish Biol. 76(7): 1863–1868.

Abbe, G.R., and D.L. Breitburg. 1992. The influence oyster toadfish (*Opsanus tau*) and crabs (*Callinectes sapidus* and Xanthidae) on survival of oyster (*Crassostrea virginica*) spat in Chesapeake Bay: Does spat protection always work? Aquaculture 107(1): 21–31.

Able, K.W., and J.A. Musick. 1976. Life history, ecology, and behavior of *Liparis inquilinus* (Pisces: Cyclopteridae) associated with sea scallops, *Plagopecten magellanicus*. Fish. Bull. U.S. 74:409–421. (red hake)

Able, K.W., M.J. Wuenschel, T.M. Grothues, J.M. Vasslides, and P.M. Rowe. 2013. Do surf zones in New Jersey provide "nursery" habitat for southern fishes? Environ. Biol. Fish 96:661–675.

Arnold, G.P., and M.G. Walker. 1992. Vertical movements of cod (*Gadus morhua* L.) in the open sea and the hydrostatic function of the swimbladder. ICES J. Mar. Sci. 49(3): 357–372.

ASMFC (Atlantic States Marine Fisheries Commission). Management section for each species, http://www.asmfc.org/.

ASMFC. 1981a. Fishery management plan for Atlantic menhaden. 146 pp. and addenda and amendments. http://www.asmfc.org/.

ASMFC. 1981b. Interstate fisheries management plan for striped bass. 329 pp. and addenda and amendments. http://www.asmfc.org/.

ASMFC. 1985a. Fishery management plan for American shad and river herring. 382 pp. and addenda and amendments. http://www.asmfc.org/.

ASMFC. 1985b. Fishery management plan for weakfish. Fishery Management Report 7. 139 pp. and addenda and amendments. http://www.asmfc.org/.

ASMFC. 1990. Fishery management plan for Atlantic sturgeon. 85 pp. and all FMP amendments and addenda. http://www.asmfc.org/.

ASMFC. 1996. Fishery management plan for tautog. Fishery Management Report 25. 69 pp. and addenda and amendments. http://www.asmfc.org/.

ASMFC. 1999. Interstate fishery management plan for American eel. 93 pp. and addenda. http://www.asmfc.org/.

ASMFC. 2002. Interstate fishery management plan for spiney dogfish. 128 pp. and all addenda. http://www.asmfc.org/.

ASMFC. 2017a. River herring stock assessment update. 682 pp. http://www.asmfc.org/.

ASMFC. 2017b.2017 Atlantic sturgeon benchmark stock assessment and peer review report. 456 pp. http://www.asmfc.org/.

ASMFC. 2019. Weakfish stock assessment update report. 93 pp. http://www.asmfc.org/.

ASMFC. 2020. American shad benchmark stock assessment and peer review report. 1188 pp. http://www.asmfc.org/.

ASMFC. 2021. Tautog regional stock assessment update 2021. 498 pp. http://www.asmfc.org/.

ASMFC. 2022. 2022 Atlantic menhaden stock assessment update. 135 pp. http://www.asmfc.org/.

Bean, T.H. 1901. Catalog of the fishes of Long Island. *In* 6th Annu. Rep. Forest, Fish and Game Comm. State N.Y., pp. 373–478.

Beguier–Pon, M., M. Castonguay, S. Shan, J. Benchetrit, and J.J. Dodson. 2015. Direct observations of American eels migrating across the continental shelf to the Sargasso Sea. Nat. Commun. 6: article 8705.

Bell, M.A., and S.A. Foster. 1994. The evolutionary biology of the threespine stickleback. Oxford University Press, New York, 571 pp.

Bell, R.J., D.E. Richardson, J.A. Hare, P.D. Lynch, and P.S. Fratantoni. 2015. Disentangling the effects of climate, abundance, and size on the distribution of marine fish: An example based on four stocks from the Northeast US Shelf. ICES J. Mar. Sci. 72(5): 1311–1322.

Bergert, B.A., and P.C. Wainwright. 1997. Morphology and kinematics of prey capture in the syngnathid fishes *Hippocampus erectus* and *Syngnathus floridae*. Marine Biol. 127:563–570.

Bigelow, H.B., and W.G. Schroeder. 1953. Fishes of the Gulf of Maine. Fish. Bull. U.S.F.W.S. 74(53), 577 pp.

Billard, R., and G. Lecointre. 2000. Biology and conservation of sturgeon and paddlefish. Rev. Fish Biol. Fish. 10:355–302.

Bisker, R.M., and M. Castagna. 1989. Biological control of crab population on hard clams *Mercenaria mercenaria* (Linnaeus,1758) by the toadfish *Opsanus tau* (Linnaeus) in tray cultures. J. Shellfish Res. 8:33–36.

Blake, R.W., K.H.S. Chan, and E.W.Y. Kwok. 2005. Finlets and the steady swimming performance of *Thunnus albacores*. J. Fish Biol. 67(5): 1434–1445.

Block, B.A., D. Booth, and F.G. Carey. 1992. Direct measurement of swimming speeds and depth of blue marlin. J. Exp. Biol. 166:267–284.

Block, B.A., H. Dewar, S.B. Blackwell, T.D. Williams, E.D. Prince, C.J. Farwell, A. Boustany, et al. 2001. Migratory movements, depth preferences, and thermal biology of Atlantic bluefin tuna. Science 293:1310–1314.

Bohlke, J.E., and C.C.G. Chaplin. 1968. Fishes of the Bahamas and adjacent tropical waters. Acad. Nat. Sci. Philadelphia. Livingston Publ., Wynnewood, PA. 771 pp.

Bolker, J.A., T.F. Hakela, and J.E. Quist. 2005. Pigmentation development, defects and patterning In summer flounder (*Paralichthys dentatus*). Zoology 108(3): 183–193.

Brainerd, E.I.. 2001. Caught in the crossflow. Nature 412:387–388. (menhaden)

Bratton, B.O., and J.L. Ayers. 1987. Observations on the electric organ discharges of two skate species (Chondrichthyes: Rajidae) and its relationship to behaviour. Environ. Biol. Fish 20(4): 241–254.

Braun, E.K., and D.P. Braun. 1994. The First Peoples of the Northeast. Moccasin Hill Press, Lincoln, MA. 144 pp.

Brewster, B. 1987. Eye migration and cranial development during flatfish metamorphosis. A reappraisal (Teleostei: Pleuronectiformes). J. Fish Biol. 31(6): 805–833. (summer flounder)

Briggs, P.T., and J.R. Waldman. 2002. Annotated list of fishes reported from the marine water of New York. Northeast. Nat. 9(1): 47–80.

Bright, M. 2000. The private life of sharks: the truth behind the myth. Stockpole Books, Mechanicsburg, PA. 320 pp. (blue shark)

Budney, L.A., and B.K. Hall, 2010, Comparative morphology and osteology of pelvic fin derival midline suckers in lumpfishes, snailfishes and gobies. J. Appl. Ichthyol. 26(2): 167–175 (lumpfish)

Burgess, G.H., B.D. Bruce, G.M. Cailliet, K.J. Goldman, R.D. Grubbs, C.G. Lowe, M.A. MacNeil, H.F Mollet, K.C. Wang, and J.B. Sullivan. 2014. A re-evaluation of the size of white shark (*Carcharodon carcharias*) population off California, USA. PLoS ONE 9(6): e98078.doi.10.10.1371/journal.pone.098078. (white shark)

Carey, F.G. 1973. Temperature regulation in free–swimming bluefin tuna. Comp. Biochem. Physiol. A 44(2): 375–392.

Carey, F.G. 1982. A brain heater in the swordfish. Science 216:1327–1329.

Carey, F.G., and B.H. Robison 1981 Daily patterns in the activities of swordfish *Xiphias gladius* observed by acoustic telemetry. Fish. Bull. U.S. 79:277–292.

Carey, F.G., and J.M. Teal. 1969. Mako and porbeagle: warm-bodied sharks. Comp. Biochem. Physiol. 28:199–204.

Carey, F.G., J.M. Teal, J.W. Kanwisher, K.D. Lawson, and J.S. Beckett. 1971. Warm-bodied fish. Am. Zool. 11:137–145. (mako)

Castro, J.L. 1983. The sharks of North American waters. Texas A&M University Press, College Station.180 pp. (spiny dogfish)

Cernadas-Masrtin, S., K.S. Rountos, J.A. Nye, M.G. Friske, and E.K. Pikitch. 2021. Composition and intraspecific variability in summer flounder (*Paralichthys dentatus*) diets in a eutrophic estuary. Front. Mar. Sci. https://doi.org/10.3389/fmars.2021.632751.

Chapple, T.K., S.J. Jorgensen, S.D. Anderson, P.E. Kanive, A.P. Klimley, L.W. Botsford, and B.A. Block. 2011. A first estimate of white shark, *Carcharodon carcharias*, abundance off Central California. Biol. Lett. 7(4): 581–583.

Cheng, C.C.-H., and A.L. DeVries. 2002. Origins and evolution of fish antifreeze proteins. *In* K.V. Ewart and C.L. Hews, eds., Fish antifreeze proteins. Molecular aspects of fish and marine biology, vol. 1, pp. 83–108. World Scientific Publ., Hackensack, NJ.

Cochran, R.C., and H.J. Grier. 1991. Regulation of sexual succession in the protogynous black sea bass, *Centropristis striatus* (Osteichthyes: Serranidae). Gen. Comp. Endocr. 82(1): 69–77.

Collette, B.B., and G. Klein-MacPhee. 2002. Bigelow and Schroeder's fishes of the Gulf of Maine, 3rd ed. Smithsonian Inst. Press, Washington, DC. 748 pp.

Collette, B.B., and C.E. Nauen. 1983. FAO species catalogue. Vol. 2. Scombrids of the world. An annotated and illustrated catalogue of tunas, mackerels, bonitos and related species known to date. FAO Fish. Synop. 125. 137 pp. (albacore).

Collins, S.P., and D. Whitehead. 2004. The functional role of passive electroreception in non–electric fishes. Anim. Biol. 54:1–25 (sharks)

Connaughton, M.A. 2004. Sound generation in the searobin (*Prionotus carolinus*) a fish with alternate sonic muscle contraction. J. Exp. Biol. 207(10): 1643–1654.

Connaughton, M.A., and M.H. Taylor. 1994. Seasonal cycles in the sonic muscles of the weakfish, *Cynoscion regalis*. Fish. Bull. 92:697–703.

Conover, D.O., and B.E. Kynard. 1981. Environmental sex determination: Interaction of temperature and genotype in a fish. Science 213:577–579. (silverside)

Curtis, T.H., C.T. McCandless, J.K. Carlson, G.B. Skomal, N.E. Kohler, L.J. Natanson, G.H. Burgess, J.J. Hoey, and H.L. Pratt Jr. 2014. Seasonal distribution and historic trends in abundance of white sharks, *Carcharodon carcharias*, in the Western North Atlantic Ocean. PloS ONE 9(6): e99240. https://doi.org/10.1371/journal.pone.0099240. (white shark)

D'Aguillo, M.C., A.S. Harold, and T.L. Darden. 2014. Diet composition and feeding ecology of the naked goby *Gobiosoma bosc* (Gobiide) from four western Atlantic estuaries. J. Fish Biol. https://doi.org/10.1111/jfb.12425.

Dickson, K.A. 1995. Unique adaptions of the metabolic biochemistry of tunas and billfishes for life in the pelagic environment. Env. Biol. Fish. 42:65–97. (little tunny)

Dijkgraaf, S. 1962. The functioning and significance of the lateral-line organs. Biol. Rev. 38:51–105. (butterfish)

Enzor, L.A., R.E. Wilborn, and W.A. Bennett. 2011. Toxicity and metabolic costs of the Atlantic stingray (*Dasyatis sabina*) venom delivery system in relation to its role in life history. J. Exp. Mar. Biol. Ecol. 409(1–2): 235–239.

Feng, A.S. 1991. Electric organs and electroreceptors. *In* C.L. Prosser, ed., Comparative animal physiology, 4th ed., pp. 317–334. Wiley, New York. (torpedo)

Fish, M.P., and W.H. Mowbray. 1970. Sounds of western North Atlantic fishes. Johns Hopkins Press, Baltimore, MD. 207 pp. (longhorn sculpin)

Fletcher, G.L, C.L. Hew, and P.L. Davies. 2001. Antifreeze proteins of teleost fishes. Ann. Rev. Physiol. 63:359–390. (Antifreeze protein essay)

Freadman, M.A. 1979. Swimming energetics of striped bass (*Morone saxatilis*) and bluefish (*Pomatomus saltatrix*): gill ventilation and swimming metabolism. J. Exp. Biol. 83:217–230. (bluefish)

Fulcher, B.A., and P.J. Motta. 2006. Suction disk performance of echeneid fishes. Can. J. Zool. 84(1): 42–50.

Garman, G.C. 1983. Observations on juvenile red hake associated with sea scallops in Frenchman Bay, Maine. Trans. Am. Fish. Soc. 112(2A): 212–215.

Gedanke, T., W.D. DuPaul, and J.A. Musick. 2005. Observations on the life history of the barndoor kkate, *Dipterus laevis*, on Georges Bank Western North Atlantic. J. Northwest Atl. Fish. Sci. 35:67–78.

Gilmore, R.G., J.W. Dodrill, and P.A. Lindley. 1983. Reproduction and embryonic development of the sand tiger shark, *Odontaspis taurus* (Rafinesque). Fish. Bull. 81(2): 201–225.

Goldman, K.J. 1997. Regulation of body temperature in the white shark, *Carcharodon carcharias*. J. Comp. Physiol. B 167:423–429.

Goode, G.B. 1884. Natural history of useful aquatic animals. Pt.3. The food fishes of the U.S. Fish. Indust. U.S. sec.1:169–549, 610–612, 629–681. (swordfish)

Gosline, W.A. 1996. Structures associated with feeding in three broad-mouthed, benthic fishes. Environ. Biol. Fishes 47:399–405. (goosefish)

Govoni, J.J., M.A. West, D. Zovotofsky, P.R. Bowser, and B.B. Collette. 2004. Ontogeny of squamation in swordfish, *Xiphias gladius*. Copeia 2004:391–396.

Greenwood, P.H., and J.R. Norman. 1975. A history of fishes. 3rd ed. John Wiley, New York. 467 pp.

Grimes, C.B., K.W. Able, and R.S. Jones. 1986. Tilefish, *Lopholatilus chamaeleonticeps*, habitat, behavior and community structure in Mid-Atlantic and southern New England waters. Environ. Biol. Fishes 15:273–292.

Grubbs, R.D, J.K. Carlson, J.G. Romine, T.H. Curtis, W.D. McElroy, C.T. McCandless, C.F. Cotton, and J.A. Musick. 2016. Critical assessment and ramifications of a purported marine trophic cascade. Sci. Rep. 6: article 20970. https://doi.org/10.1038/srep20970. (cownose ray)

Hagan, S.M., S.A. Brown, and K.W. Able. 2007. Production of mummichog (*Fundulus heteroclitus*): Response in marshes treated for common reed (*Phragmites australis*) removal. Wetlands 27:54–67.

Hare, J.A., J.H. Churchill, R.K. Cowen, T.J. Berger, P.C. Cornillon, P. Dragos, S.M. Glenn, J.J. Govoni and T.N. Lee 2002 Routes and rates of larval fish transport from the Southeast to the Northeast United States continental shelf. Limnol. Oceanogr. 47(6): 1774–1789. (cornetfish)

Helfield, J.M., and R.J. Naiman. 2001. Effects of salmon-derived nitrogen on riparian forest growth and implications for stream productivity. Ecology 82(9): 2403–2409. (anadromous alewife)

Hildebrand, S.F., and W.C. Schroeder. 1928. Fishes of Chesapeake Bay. Bull. U.S. Bur. Fish., vol. 43, pt. 1. 388 pp.

Hoogland, R., D. Morris, and N. Tinbergen. 1956. The spines of sticklebacks

(*Gasteroseus* and *Pygosteus*) as means of defence against predators (*Perca* and *Esox*). Behaviour 10(3): 205–236.

Howell, P., and P.J. Auster. 2012. Phase shift in an estuarine finfish community associated with warming temperatures. Mar. Coast. Fish. 4(1): 481–495.

Hungerford, J.M. 2010. Scombroid poisoning: A review. Toxicon 56(2): 231–243.

Jessop, B.M. 1994. Homing of alewives (*Alosa pseudoharengus*) and blueback herring (*A. aestivalis*) to and within Saint John River, New Brunswick, as indicated by tagging data. Can. Tech. Rep. Fish. Aquat. Sci. 2015. 22 pp.

Jordan, D.S., and B.W. Evermann. 1896–1900. The fishes of North and Middle America. A descriptive catalogue of the species of fish-like vertebrates. Bull. U.S. Natl. Mus. 47, pts. 1–4. 3313 pp. + 392 plates.

Joseph, J., W. Klawe, and P. Murphy. 1979. Tuna and billfish—fish without a country. Inter-American Tropical Tuna Commission, San Diego, CA. 46 pp. (scombrids)

Juanes, F., and D.O. Conover. 1995. Size-structured piscivory advection and the linkage between predator and prey recruitment in young-of-the-year bluefish. Mar. Ecol. Prog. Ser. 128:287–304.

Kaijura, S.M., J.B. Forni, and A.P. Summers. 2005. Olfactory morphology of carcharhinid sharks: Does the cephalofoil confer a sensory advantage? J. Morph. 264(3): 253–263. (hammerhead shark)

Kilduff, P., J. Carmichel, and R. Latour. 2009. *In* T. Berger, ed., Guide to fisheries science and stock assessments. Atlantic States Marine Fisheries Commission, Arlington, VA. 66 pp.

Kim, S.H., K. Shimada, and C.K. Rigsby. 2013. Anatomy and evolution of heterocercal tail in lamniform sharks. Anat. Rec. 296:433–442. (thresher shark)

Kissil, G.W. 1974. Spawning of the anadromous alewife, *Alosa pseudoharengus* in Bride Lake, Connecticut. Trans. Am. Fish. Soc. 103:312–317.

Kleckner, R.C. 1980. Swim bladder volume maintenance related to initial oceanic migratory depth in silver-phase *Anguilla rostrata*. Science 208:1481–1482.

Klimley, A.P. 1999 Sharks beware. Am. Sci. 87:488–491. (white shark)

Klimley, A.P. 2013 The biology of sharks and rays. University of Chicago Press, Chicago. 528 pp. (white shark)

Kotrschal, K. 1996. Solitary chemosensory cells: Why do primary aquatic vertebrates need another taste system? Trends Ecol. Evol. 11(3): 110–114. (searobin)

Lavende, N. 1949. Sexual differences and normal protogynous hermaphroditism in the Atlantic sea bass, *Centropristis striatus*. Copeia 1949(3): 185–194.

Leggett, W.C., and J.E. Carscadden. 1978. Latitudinal variation in reproductive characteristics of American shad *Alosa sapidissima*: Evidence for population specific life history strategies in fish. J. Fish. Res. Bd. Can. 35:1469–1478.

Li, W., A.P. Scott, M.J. Siefkes, H. Yan, Q. Liu, S. Yun, and D. Gage. 2002. Bile acid secreted by male sea lamprey that act as a sex pheromone. Science 296:138–141.

Liem, K.F., and S.L. Sanderson. 1986. The pharyngeal jaw apparatus of labrid fishes. A functional morphological perspective. J. Morph. 187(2): 143–158.

Lindgren, M., O. Ostman, and A. Gardmark. 2011. Interacting trophic forcing and the population dynamics of herring. Ecology 92(7): 1407–1413.

Loesch, J.G. 1987. Overview of life history aspects of anadromous alewife and blueback herring in freshwater habitats. *In* M.J. Dadswell et al., eds., Common strategies of anadromous and catadromous fishes. Am. Fish. Soc. Symp. 1:89–103.

MAFMC (Mid–Atlantic Fishery Management Council). 1978. Fishery management plan for the Atlantic mackerel fishery of the Northwest Atlantic Ocean. 129 pp. and amendments and frameworks.

MAFMC. 1987. Fishery management plan for the summer flounder fishery. 157 pp. and amendments, frameworks, and addenda. https://www.mafmc.org/.

MAFMC. 1990. Fishery management plan for the bluefish fishery, and amendments and frameworks. https://www.mafmc.org/.

MAFMC. 1996. Amendment 8 to the summer flounder fishery management plan: Fishery management plan for the scup fishery 87 pp. and addenda and amendments. https://www.mafmc.org/.

MAFMC. 1999. Spiney dogfish fishery management plan and amendments. https://www.mafmc.org/.

MAFMC. 2018. Update on the status of spiney dogfish in 2018. https://static1.squarespace.com/static/511cdc7fe4b00307a2628ac6/t/5d2390c2aee3e800017d0b69/1562611910802/2018+Status+Report+for+spiny+dogfish.pdf.

MAFMC. 2000. Tilefish fishery management plan. 439 pp. and amendments and frameworks. https://www.mafmc.org/.

Magnuson, J.J., C. Safina, and M.P. Sissenwsine. 2001. Whose fish are they anyway? Science 293:1267–1268. (scombrids)

Malins, D.C., and A. Barone 1970 Glyceryl ether metabolism: Regulation of buoyancy in dogfish *Squalus acanthias*. Science 167:79–80.

Marshall, W.S., and M. Groswell. 2006. Ion transport, osmoregulation, and acid–base balance. *In* D.H. Evans and J.D.Clairborne, eds., The physiology of fishes, 3rd ed. pp.177 230. CRC Press, Boca Raton, Fl.. (sharks)

Martin, F.D. 1972. Factors influencing the local distribution of *Cyprinodon variegatus* (Pisces: Cyprinodontidae). Trans. Am. Fish. Soc. 101:89–93.

Mast, S.O. 1916. Changes in shade, color and pattern in fishes, and their bearing on the problems of adaptation and behavior with especial reference to the flounders, *Paralichthys* and *Ancyclopsetta*. Bull. U.S. Bur. Fish. 34:177–238.

McBride, R.S., and K.W, Able. 1998. Ecology and fate of butterflyfishes, *Chaetodon* spp., in the temperate, Western North Atlantic. Bull. Mar. Sci. 63(2): 401–416.

McCartin, K, A. Jordaan, M. Sclafani, R. Cerrato, and M.G. Fisk. 2019. A new paradigm in Alewife migration:Ooscillations between spawning grounds and estuarine habitats. Trans. Am. Fish. Soc. 148 (3): 605–619.

McComb, D.M., T.C. Trincas, and S.M. Kajiura. 2009. Enhanced visual fields in hammerhead sharks. J. Exp. Biol. 212:4010–4018.

McDowall, R.M. 1981. A sub-dorsal fin pore/canal system in the centrolophid fish *Schedophilus maculatus* (Pisces: Stromateoidei). Copeia 1981(2): 492–494. (butterfish)

McGowan, C. 1988. Differential development of the rostrum and mandible of the swordfish (*Xiphias gladius*) during ontogeny and its possible functional significance. Can. J. Zool. 66:496–503.

McHugh, J.L., and A. Williams. 1976. Historical statistics of the New York Bight area. N.Y. Sea Grant Institute, State University of New York–Stony Brook. 73 pp.

McKeown, B.A. 1984. Fish migration. Croom Helm, London. 224 pp. *(Clupea harengus)*

Meyer, T.L., R.A. Cooper, and R.W. Langton 1979 Relative abundance, behavior, and food habits of the American sand lance, *Ammodytes americanus*, from the Gulf of Maine. Fish. Bull. U.S. 77:243–253.

Miyake, M., and S. Hayasi. 1972. Field manual for statistics and sampling of Atlantic tunas and tuna-like fishes. International Commission for the Conservation of Atlantic Tunas, Madrid.

Montgomery, J.C., C.F. Baker, and A.G. Carton 1997 The lateral line can mediate rheotaxis in fish. Nature 389:960–963. (pollock)

Moore, K.S., S. Wehrli, H. Roder, M. Rogers, J.N. Forrest Jr., D. McCrimmon, and M. Zaasloff. 1993. Squalamines: An aminosterol antibiotic from the shark. PNAS 90(4): 1354–1358. (spiny dogfish)

Myers, R.A., J.K. Baum, T.D. Shepherd, S.P. Powers, and C.H. Peterson. 2007. Cascading effects of the loss of apex predatory sharks from the coastal ocean. Science 315:1846–1850. (cownose ray)

Nammack, M.F., J.A. Musick, and J.A. Colvocoresses. 1985. Life history of spiny dogfish of the northeastern United States. Trans. Am. Fish. Soc. 114(3): 366–376.

Natanson, L.J., and G.B. Skomal. 2015. Age and growth of the white shark, *Carcharaodon carcharias*, in the western North Atlantic Ocean. Mar. Freshw. Res. 66(5): 387–398.

Neudecker, S. 1989. Eye camouflage and fake eyespots: Chaetodontid responses to predators. Envir. Biol. Fish. 25(1–3): 143–157.

NEFMC (New England Fishery Management Council). 1985. Fishery management plan for the Northeast multi-species fishery. 558 pp. and amendments and frameworks. https://www.nefmc.org/.

NEFMC. 1998. Monkfish fishery management plan. 405 pp. and amendments and frameworks. https://www.nefmc.org/.

NEFMC. 2020. Stock assessment and fishery evaluation (SAFE report) for the small-mesh multispecies fishery, fishing years 2017–2019. 45 pp. https://www.nefmc.org/.

NMFS. 2006. Final consolidated Atlantic highly migratory species fishery management plan. 1629 pp. and amendments. https://media.fisheries.noaa.gov/dam-migration/atlantic-hms-consolidated-fmp.pdf.

NMFS. 2016. NOAA fisheries navigator. Commercial Fisheries News, Special Suppl., March 2016. 8pp.

NMFS. 2019. 66th Northeast regional stock assessment workshop (66th SAW) assessment report. NEFSC (Northeast Fisheries Science Center) Ref. Doc. 19–08. 733 pp.

NMFS. 2020. Operational assessment of the black sea bass, scup, bluefish and monkfish stocks, updated through 2018. NEFSC Ref. Doc. 20–01.164 pp.

NMFS. 2022a. Management track assessments fall, 2021. NEFSC Ref. Doc.22-07. 42pp.

NMFS. 2022b. 2021 Management track assessment of black sea bass, golden tilefish, scup and Atlantic mackerel. NEFSC Ref. Doc. 22–10.104 pp.

Nordie, F.G. 1985. Osmotic regulation in the sheepshead minnow *Cyprinodon variegatus* Lacepede. J. Fish. Biol. 26(2): 161–170.

Oliveira, J.S., O.R. Pires Jr., R.A.V. Morales, C. Bloch, C.A. Schwartz, and J.D. De Freitas. 2003. Toxicology of pufferfish: two species (*Lagocephalus laevigatus*, Linnaeus 1766 and *Sphoeroides spengleri*, Bloch 1785) from the Southeastern Brazilian coast. J. Venom. Anim. Toxins Incl. Trop. Dis. 9:76–88. (bandtail puffer)

Olla, B.L., A.J. Bejda, and A.D. Martin. 1974. Daily activity, movements, feedings, and seasonal occurrence in the tautog, *Tautoga onitis*. Fish. Bull. U.S. 72:27–35,

Pelster, B., and P. Scheid. 1993. Metabolism of the swimbladder epithelium and the single concentrating effect. Comp. Biochem. Physiol A 105(3): 383–388. (cod)

Petersen, J.C., and J.B. Ramsay. 2020. Walking in chains: The morphology and mechanics behind the fin ray derived limbs of sea-robins. J. Exp. Biol. 223(18): 227140

Petzel, D.H., H.M. Reisman, and A.L. DeVries. 1980. Seasonal variation of antifreeze peptide in winter flounder, *Pseudopleuronectes americanus*. J. Exp. Zool. 211:63–69.

Pitcher, T.J., B.L. Partridge, and C.S. Wardle. 1976. A blind fish can school. Science 194:963–965. (pollock)

Reisman, H.M. 1968. Reproductive isolating mechanisms of the blackspotted stickleback, *Gasterosteus wheatlandi*. J. Fish. Res. Bd. Can. 25(2): 2703–2706.

Reisman, H.M., G.L. Fletcher, M.H. Kao, and M.A. Shears. 1987. Antifreeze proteins in the grubby sculpin, *Myoxocephalus aeneus* and the tomcod, *Microgadus tomcod*: comparisons of seasonal cycles. Environ. Biol. Fishes 18:295–301. (grubby)

Reisman, H.M., M.H. Kao, and G.L. Fletcher. 1984. Antifreeze glycoprotein in a "southern" population of Atlantic tomcod, Microgadus tomcod. Comp.Biochem. Physiol. A 78:445–447.

Ripley, J.L., and C.M. Foran. 2009. Direct evidence for embryonic uptake of paternally-derived nutrients in two pipefishes (*Syngnathidae: Syngnathus* spp.). J. Comp. Physiol. B 179:325–333.

Roberts, J.L. 1975. Active branchial and ram gill ventilation in fishes. Biol. Bull. 148:85–105. (bluefish)

Roberts, J.L. 1978. Ram gill ventilation in fish. *In* G.D. Sharp and A.E. Dizon, eds., The physiological ecology of tunas, pp.83–88. Academic Press, New York.

Rome, L.C., and S.L. Lindstedt. 1998. The quest for speed: Muscles built for high-frequency contractions. New Physiol. Sci. 13:261–268. (toadfish)

Rome, L.C., D. Syme, S. Hollingsworth, S. Lindstedt, and S. Baylor. 1996. The whistle and the rattle: The design of sound producing muscles. PNAS 93:8095–8100. (toadfish)

Rowe, S., and J.A. Hutchings. 2004. The function of sound production by Atlantic cod as inferred from patterns of variation in drumming muscle mass. Can. J. Zool. 82(9): 1391–1398.

Safina, C., and J. Burger. 1985. Common tern foraging: Seasonal trends in prey fish densities and competition with bluefish. Ecology 66:1457–1463.

SAFMC (South Atlantic Fishery Management Council). 2003. Fishery management plan for the dolphin and wahoo fishery of the Atlantic. 309 pp. and amendments. https://safmc.net/.

Sanderson, S.L., A.Y. Cheer, J.S. Goodrich, J.D. Graziano, and W.T. Callan 2001 Crossflow filtration in suspension-feeding fishes. Nature 412:439–441. (menhaden)

Saunders, M.W., and G.A. McFarlane 1993 Age and length at maturity of the female spiny dogfish, *Squalus acanthias*, in the Strait of Georgia, British Columbia, Can. Env. Biol. Fish. 38(1–3): 49–57.

Savaria, M.C., and N.J. O'Connor. 2013. Predation of the non-native Asian shore crab *Hemigrapsus sanguineus* by a nativ fish species, the cunner (*Tautogolabrus adspersus*). J. exp. Mar. Biol. Ecol. 449:335–339.

Schwab, I.R. 2004. If looks could kill. J. Ophthalmol. 88(12): 1486. https://doi.org /1136/bjo.2004.057232. (northern stargazer)

Schwartz, F.J. 1997. Biology of the striped cusk eel, *Ophidion marginatum*, from North Carolina. Bull. Mar. Sci. 61:327–342.

Shadwick, R.E., S.L. Katz, K.E. Korsmeyer, T. Knower, and J.W. Covell. 1999. Muscle dynamics in skipjack tuna: Timing of red muscle shortening in relation to activation and body curvature during steady swimming. J. Exp. Biol. 202:2139–2150.

Shiffman, D.S., and N. Hammerschlag 2016. Shark conservation and management policy: a review and primer for non-specialists. Animal Conservation 19(5): 401–412.

Sims, D.W., and V.A. Quayle. 1998. Selective foraging behavior of basking sharks on zooplankton in a small-scale front. Nature 1393:460–464.

Smith, C.F., and E. Hasbrouck. 1988. A guide to identifying tuna in New York area waters. Cornell Coop. Ext. Suffolk Co. Mar. Prog., Riverhead, NY. 20pp.

Strong, J.A. 1997. The Algonquian Peoples of Long Island from earliest times to 1700. Empire State Books, Interlaken, NY. 368 pp.

Summers, A., and T. Koob 1996. On the hydrodynamic shape of little skate (*Raja erinacea*) egg capsules. Bull. Mt. Desert Isl. Biol. Lab. 35:108–111.

Thorrold, S.R., C. Latkoczy, P. Swart, and C.M. Jones. 2001. Natal homing in a marine fish metapopulation. Science 291:297–299. (weakfish)

Thunberg, B.E. 1971. Olfaction in parent stream selection in the alewife (*Alosa pseudoharengus*). Anim. Behav. 19(2): 217–225.

Wainwright, P.C., R.G. Turingan, and E.L. Brainerd. 1995. Functional morphology of pufferfish inflation: Mechanism of the buccal pump. Copeia 1995:614–625. (porcupinefish)

Waldman, J.R., J. Grossfield, and I. Wirgin. 1988. Review of stock discrimination techniques for striped bass. N. Am. J. Fish Manag. 8(4): 410–425.

Walker, B.W. 1959. A guide to the grunion. Calif. Fish Game 38(3): 409–420.

Wegner, N.C., O.E. Snodgrass, H. Dewar, and J.R. Hyde. 2015. Whole-body endothermy in a mesopelagic fish, the opah, *Lampris guttatus*. Science 348(6238): 786–789. (swordfish)

Weis, J.S. 2002. Tolerance to environmental contaminants in the mummichog, *Fundulus heteroclitus*. Hum. Ecol. Risk Assess.: Intern. J. 8(5): 933–953.

Weis, J.S., and P. Weis. 1977. The effects of heavy metals on embryonic development of the killifish *Fundulus heteroclitus*. J. Fish. Biol. 11:49–54.

Whitfield, P.E., T. Gardner, S.P. Vives, M.R. Gilligan, W.R. Courtenay Jr., G.C. Ray, and J.A. Hare. 2002. Biological invasion of the Indo-Pacific lionfish *Pterois volitans* along the Atlantic coast of North America. Mar. Ecol. Progr. Ser. 235:289–297.

Wilbur, C.K. 1996. The New England Indians. 2nd ed. Globe Pequot Press, Guilford, CT. 123 pp.

Wilson, A.B., I. Ahnesjo, A.C.J. Vincent, and A. Meyer. 2003. The dynamics of male brooding, mating patterns, and sex roles in pipefishes and seahorses (Family Syngnathidae). Evolution 57(6): 1374–1386.

Wood, A.J.M., J.S. Collie, and J.A. Hare. 2009. A comparison between warm-water fish assemblages of Narragansett Bay and those of Long Island Sound. Fish. Bull 107:89–100.

Wooton, R.J. 1984. A functional biology of sticklebacks. University of California Press, Berkeley. 265 pp.

ADDITIONAL SOURCES

History of New York's Marine Fishes Reports

One of the earliest contributions was "The Fishes of New York, described and arranged" by Samuel Latham Mitchill (Trans. Lit. Phil. Soc.1814(1): 355–492). This article included 147 species, 60 of them illustrated, and with only a few exceptions these were marine fishes.

Another early work was by James E. DeKay (1842), *Zoology of New York. IV. Fishes* (Nat. Hist. N.Y. Geol. Surv., 415 pp.). DeKay listed 355 marine, estuarine, and freshwater species. However, because he occasionally counted different forms of the same fish as separate species, his total number of nearly 200 marine and estuarine species was somewhat inflated. Between 1842 and 1844, William O. Ayres contributed to the study of Long Island fishes in his "Enumeration of the Fishes of Brookhaven, Long Island" (Boston J. Nat. Hist. 4:255–392).

In 1901 Tarleton H. Bean authored a "Catalog of the Fishes of Long Island" (6th Annu. Rep. Forest, Fish, Game Comm. State N.Y., pp. 373–478). In 1903, he published a *Catalogue of the Fishes of New York* (Bull. N.Y. State Mus. 60. 784 pp.). In that latter report, approximately 200 of the 360 annotated species were marine forms.

Roy Latham, a Long Island naturalist, published a series of reports in *Copeia* (1916–1923) recording the fishes observed on eastern Long Island at Orient, New York.

Although there were many small regional studies of fishes within the state during the first part of the twentieth century, it was not until the New York State Conservation Department initiated a complete biological survey of the state's fresh and salt waters that New York was the object of systematic studies. The following contributions were of particular relevance:

Dickinson, C.M. 1939. Commercial fisheries. *In* A biological survey of the salt waters of Long Island. 28th Ann. Rept. N.Y. State Conserv. Dept. 1938(Suppl.): 15–16.

Greeley, J.R. 1939. Fishes and habitat conditions of the shore zone based upon July and August seining investigations. *In* A biological survey of the salt waters of Long Island. 28th Ann. Rep. N.Y. State Conserv. Dept. 1938(Suppl.): 73–91.

Perlmutter, A. 1939. An ecological survey of young fish and eggs identified from two-net collections. *In* A biological survey of the salt waters of Long Island. 28th. Ann. Rep. N.Y. State Conserv. Dept. 1938(Suppl.): 11–71.

More recent collections of marine fishes new to New York and results of faunal studies in particular bay and shoreline locations are as follows:

Alperin, I.M., and R.H. Schaefer. 1965. Marine fishes new or uncommon to Long Island, New York. N.Y. Fish Game J. 12(1): 1–16.

Briggs, P.T., and J.S. O'Connor. 1971. Comparisons of shore-zone fishes over naturally vegetated and sand-filled bottoms in Great South Bay. N.Y. Fish Game J. 18(1): 15–41.

Briggs, P.T., and J.R. Waldman. 2002. Annotated list of fishes reported from the marine water of New York. Northeast. Nat. 9(1):47–80. (This is the most recent and valuable list of marine fishes found on Long Island. Our treatment is not as exhaustive as this. We do not include many species that are exclusively pelagic, deep sea, or freshwater strays.)

Hickey, C.R., Jr. 1985. Survey of the technical literature on the marine finfishery resources of the Peconic/Gardiners Bay system, New York, 1900–1984, with recommendations for resource conservation and study. Marine Science Research Center Living Marine Resources Institute, State University of New York–Stony Brook, Spec. Rept. 65. 106 pp.

Wood, A.J.M., J.S. Collie, and J.A. Hare. 2009. A comparison between warm-water fish assemblages of Narragansett Bay and those of Long Island Sound. Fish. Bull. 107:89–100.

Wright, J.J., R.E. Schmidt, and B.R Weatherwax. 2016. New and previously overlooked records of several fish species from the marine waters of New York. Northeast. Nat. 23(1):118–133.

Works about Resident and Seasonal Species Found in New York Marine Waters

Able, K.W., and M.P. Fahay. 2010. Ecology of estuarine fishes: temperate waters of western North Atlantic. Johns Hopkins University Press, Baltimore, MD. 566 pp.

Bigelow, H.B., and W.C. Schroeder. 1953. Fishes of the Gulf of Maine. Fish. Bull. U.S. 53:1–577.

Bohlke, J.E., and C.C.G. Chaplin. 1968. Fishes of the Bahamas and adjacent tropical waters. Acad. Nat. Sci. Philadelphia, Livingston Publ., Wynnewood, PA. 771 pp.

Breder, C.M., Jr. 1948. Field book of marine fishes of the Atlantic coast. Putnam, New York. 322 pp.

Clayton, G., C. Cole, S. Murawski, and J. Parrish. 1976. Common marine fishes of coastal Massachusetts. Mass. Coop. Fish. Res. Unit, Contrib. 54. 231 pp.

Coad, B.W. 1992. Guide to the marine sport fishes of Atlantic Canada and New England. University of Toronto Press, Toronto. 307 pp.

Collette, B.B., and G. Klein–MacPhee. 2002. Bigelow and Schroeder's fishes of the Gulf of Maine. 3rd ed. Smithsonian Institution Press, Washington, DC. 748 pp. (The latest edition of the most useful source for detailed biological information regarding most of the marine fishes seen in and around New York.)

DeMaddalena, A., and W. Heim. 2010. Sharks of New England. Down East Books, Rockport, Maine. 183 pp.

Fowler, H.W. 1905. The fishes of New Jersey. Rep. N.J. State Mus. 1905(1906): 35–477.

Gordon, B.L. 1960. Marine fishes of Rhode Island. Book and Tackle Shop, Watch Hill, Rhode Island. 136 pp.

Grosslein, M.D., and T.R. Azarovitz. 1982. Fish distribution. New York Sea Grant Institute, Marine Ecosystems Analysis Program, New York Bight Atlas Monograph 15, Albany. 182 pp.

Hoesse, H.D., and R.H. Moore 1998. Fishes of the Gulf of Mexico. 2nd ed. Texas A&M University Press, College Station. 422 pp.

Humann, P. 1989. Reef fish identification Florida, Caribbean, Bahamas. New World Publ., Jacksonville, FL. 288 pp.

Kells, V., and K. Carpenter 2011. A field guide to coastal fishes from Maine to Texas. Johns Hopkins University Press, Baltimore, MD. 446 pp.

Lieske, E., and R. Myers. 1966. Coral reef fishes Caribbean, Indian Ocean and Pacific Ocean. Princeton University Press, Princeton, NJ. 400 pp.

Lythgoe, J., and G. Lythgoe. 1992. Fishes of the sea. The North Atlantic and Mediterranean. MIT Press, Cambridge, MA. 256 pp.

McClane, A.J., ed. 1974. Field guide to saltwater fishes of North America. Holt, Rinehart Winston, New York. 283 pp.

McEachran, J.D., and J.D. Fechhelm. 1998. Gulf of Mexico. Vol 1, Myxiniformes–Gasterosteiformes. University of Texas Press, Austin. 1120 pp.

McEachran, J.D., and J.D. Fechhelm. 2005. Gulf of Mexico. Vol 2, Scorpaeniformes–Tetraodontiformes. University of Texas Press, Austin. 1014 pp.

Murdy, E.O., R.S. Birdsong, and J.A. Musick. 1997. Fishes of Chesapeake Bay. Smithsonian Institution Press, Washington, DC. 324 pp.

Murdy, E.O., and J.A. Musick. 2013. Field guide to fishes of the Chesapeake Bay. Johns Hopkins University Press, Baltimore, MD. 345 pp.

Nichols, J.T. 1918. Fishes of the vicinity of New York City. Am. Mus. Nat. Hist. Handbook Ser. 7. 118 pp.

Nichols, J.T., and C.M. Breder Jr. 1927. The marine fishes of New York and southern New England. Zoologica 9:1–192.

Randall, J.E. 1983 Caribbean reef fishes. 2nd ed. TFH Publ., Neptune City, NJ. 320 pp.

Roberts, M.F. 1985. The tidemarsh guide to fishes. Saybrook Press, Old Saybrook, CT. 373 pp.

Robins, C.R., G.C. Ray, and J. Douglass. 1986. A field guide to Atlantic coast fishes of North America. Houghton Mifflin, Boston. 354 pp.

Scott, W.B., and M.G. Scott. 1988. Atlantic fishes of Canada. Can. Bull. Fish. Aquat. Sci. 219. 731 pp.

Smith-Vaniz, W.F., B.B. Collette, and B.E. Luckhurst 1999 Fishes of Bermuda: History, zoogeography, annotated checklist, and identification keys. ASIH Spec. Publ. 4. 424 pp.

Thompson, K.S., W.H. Weed III, A.G. Taruski, and D.E. Simanek. 1978. Saltwater fishes of Connecticut. 2nd ed. State Geol. Nat. Hist. Surv. Conn. Bull. 105, Hartford. 186 pp.

Ursin, M. 1977. A guide to fishes of the temperate Atlantic coast. Dutton, New York. 262 pp.

Waldman, J. 2013. Running silver: Restoring Atlantic rivers and their great fish migrations. Lyons Press, Guilford, CT. 284 pp.

Weiss, H.M. 1995. Marine animals of southern New England and New York. State Geol. Nat. Hist. Surv. Conn. Dept. Env. Protection Bull 115, Hartford.

A Selection of General Fish Books and Textbooks

Barton, M. 2007. Bond's biology of fishes. 3rd ed. Thompson Brooks/Cole, Belmont, CA. 891 pp.

Bone, Q., N.B. Marshall, and J.H.S. Blaxter 1995. Biology of fishes. 2nd ed. Blakie Academic and Professional, London. 332 pp.

Gordon, B.L. 1977. The secret lives of fishes. Grosset Dunlap, New York. 305 pp.

Greenwood, P.H., and J.R. Norman. 1975. A history of fishes. 3rd ed. John Wiley, New York. 467 pp.

Hastings, P.A., H.J. Walker Jr., and G.R. Grantly. 2014. Fishes: A guide to their diversity. University of California Press, Oakland. 311 pp.

Helfman, G.S., B.B. Collette, D.E. Facey, and B.W. Bowen. 2009. The diversity of fishes. 2nd ed. Wiley-Blackwell, Oxford. 736 pp.

Klimley, A.P. 2013. The biology of sharks and rays. University of Chicago Press, Chicago. 528 pp.

Marshall, N.B. 1966. The life of fishes. World Publ., Cleveland, OH. 402 pp.

Moyle, P.B., and J.J. Cech Jr. 2004. Fishes: An introduction to ichthyology. 5th ed. Prentice Hall, Upper Saddle River, NJ. 726 pp.

Payton, J.R., and W.N. Eschmeyer, eds. 1995. Encyclopedia of fishes. Academic Press, San Diego. 240 pp.

Special Resources

Gregory, W.K. 1933. Fish skulls: A study of the evolution of natural mechanisms. Trans. Am. Philos. Soc. New Ser. 23:75–481.

Jaeger, E.C. 1978. A source-book of biological names and terms. 3rd ed. Charles C. Thomas Publ., Springfield, IL. 360 pp. (A source of Latin and Greek origins of scientific names.)

Jordan, D.S., and B.W. Evermann. 1896–1900. The Fishes of North and Middle America. A descriptive catalogue of the species of fish-like vertebrates. Bull. U.S. Nat. Mus. 47, pts. 1–4. 3313 pp. + 392 plates. (One of the most influential fish studies every published, this four-part set by David Starr Jordan and Barton Warren Evermann served as the basis of many of the early twentieth century regional studies and is routinely cited as a valuable historical resource. Jordan and Evemann's volumes were the authoritative source of classification, synonyms, and identification keys for 227 families, 1077 genera, and 3127 fish species. The material included extensive anatomical descriptions that distinguished each of those groups and a species' geographic distribution. We used this work as a source for the Greek and Latin origin of a species' scientific name. One of us (HMR) acquired a copy of Jordan and Evermann while serving as a postdoctoral fellow at Cornell University. During one of many visits to the Albert R. Mann Library on campus, the library was disposing of excess and unwanted books. Remarkably, among them was this set. An identification label on the inside cover revealed that this set was once owned by Andrew Dickson White who, along with Ezra Cornell, cofounded the university.)

McHugh, J.L. 1972. Marine fisheries of New York State. Fish. Bull. 70(1): 585–610.

McHugh, J.L., and E. Hasbrouck. 1990. Fishery management in New York Bight: Experience under the Magnuson Act. Fish. Res. 8:205–221.

Miyake, M., and S. Hayasi. 1972. Field manual for statistics and sampling of Atlantic tunas and tuna-like fishes. International Commission for the Conservation of Atlantic Tunas, Madrid.

Nelson, J.S. 2006. Fishes of the world. 4th ed. John Wiley, Hoboken, NJ. 601 pp. (The most recent edition of a systematic treatment of all major fish groups including those anatomical features that distinguish the 62 orders and 515 families of the world's fishes. Further information includes the general world distribution of each family and the number of genera and species within each family.)

Page, L.M., H. Espinosa-Perez, L.T. Findley, C.R. Gilbert, R.N. Lea, N.E. Mandrak, R.L. Mayden, and J.S. Nelson. 2013. Common and scientific names of fishes from the United States, Canada, and Mexico. 7th ed. Am. Fish. Soc. Special Publ. 34. 384 pp. (The most widely recognized source of bona fide common names for North American fish species.)

Online Resources for Biological and Fisheries Information

Atlantic States Marine Fisheries Commission
httpы//asmfc.org

The Atlantic States Marine Fisheries Commission helps to sustainably and cooperatively manage coastal fishery resources and addresses fisheries management, fisheries science, and habitat.

Choose Local F.I.S.H. (Fresh. Indigenous. Sustainable. Healthy)
www.localfish.org/

Cornell Cooperative Extension, Suffolk County—Fisheries Department
https://ccesuffolk.org/marine/fisheries

Cornell Cooperative Extension Fisheries is a department within Cornell Cooperative Extension of Suffolk County's Marine Program. The Fisheries Department focuses on all things fisheries—testing new and upcoming fishing gear, developing conservation gear technology, collecting biological data on the species most important to our fishing industries, holding events to introduce local seafood to Long Islanders, and so much more. The Fisheries Department is here to ensure the voice of the local marine fishing industry is heard, whether it be for new fishing regulations or coexisting with ocean construction projects.

The Elasmodiver Shark and Ray Field Guide

https://elasmodiver.com

> Elasmobranchs (sharks, skates, and rays) are featured in the form of a field guide with photos and a brief profile of 148 elasmobranchs.

Encyclopedia of Life

https://eol.org/info/fishes

> The Encyclopedia of Life includes information about all known animal and plant species. Details vary but many species include photos.

FishBase

https://fishbase.se

> FishBase is a global information system on fishes and includes details of the world's fishes. Further, it includes a computer-based course in ichthyology for upper undergraduate and graduate students in biology and environmental science.

Fish Guy Photos

https://www.fishguyphotos.com

> Chris Paparo is a wildlife photographer, writer, and lecturer. His interest in coastal ecology, fishing, and the outdoors led him to obtain a B.S. in Marine Science from Southampton College of Long Island University, and he is currently the manager of the Marine Science Center at Stony Brook University, Southampton.
>
> In addition to freelance writing for several fishing and wildlife related publications, Paparo currently writes the monthly "Naturalist's Logbook" column for the New York–New Jersey edition of *On the Water Magazine*. Although his work tends to focus on marine life, everything in the natural world is fair game. Keep up with Paparo's adventures by following him on
> • Facebook • https://www.facebook.com/fishguyphotos
> • Twitter • https://twitter.com/fishguyphotos
> • Instagram • https://www.instagram.com/fishguyphotos/

Florida Museum of Natural History

https://www.floridamuseum.ufl.edu/discover-fish/species-profiles/

> The Florida Museum of Natural History of Gainesville, Florida, is one of the premier museums presenting information on marine biology. The site includes more than 260 biological profiles of fish species, some of them observed in New York waters.

Food and Agriculture Organization of the United Nations: Fisheries and Aquaculture

https://www.fao.org/fishery/en

> The Food and Agriculture Organization of the United Nations addresses fisheries statistics, information, and technology.

International Game Fish Association

https://igfa.org/igfa–world–records–search/

Marine Laboratory and Aquarium

https://mote.org

> The Mote Marine Laboratory of Sarasota, Florida, emphasizes research with an emphasis on sharks. Many of those occur in New York waters.

Mid-Atlantic Fishery Management Council

https://www.mafmc.org

The Mid-Atlantic Fishery Management Council develops fishery management plans and recommends management measures to ensure the long-term sustainability of Mid-Atlantic fisheries.

National Marine Fisheries Service

https://www.greateratlantic.fisheries.noaa.gov

The National Marine Fisheries Service is a subsidiary of the National Oceanic and Atmospheric Administration (Department of Commerce). The NMFS is responsible for the stewardship of the nation's ocean resources and their habitat. Programs of the NMFS include aquaculture, habitat conservation, protected species, fisheries statistics, stock assessments, and sustainable fisheries.

National Oceanic and Atmospheric Administration

https://www.fisheries.noaa.gov

Fisheries Office of Science and Technology. Commercial landings query. Available at https://www.fisheries.noaa.gov/foss.

National Marine Fisheries Service (NMFS), Fisheries Statistics Division. Personal communication recreational landings. Available at https://www.fisheries.noaa.gov/foss/.

NMFS Greater Atlantic Regional Fisheries Office. Regulations section for each species. Available at https://www.fisheries.noaa.gov/new–england–mid –atlantic/resources–fishing/resources–fishing–greater–atlantic–region.

New England Fishery Management Council

https://www.nefmc.org

The New England Fishery Management Council develops fishery management plans and recommends management measures to ensure the long term sustainability of New England fisheries.

New York Sea Grant

https://seagrant.sunysb.edu/

Philip Briggs, a NYSDEC fisheries biologist, constructed an inshore saltwater fish calendar for New York showing those months of the year when fishing was good or excellent for the following species: striped bass, bluefish, snapper blues, winter flounder, fluke (summer flounder), weakfish, porgy (scup), blackfish (tautog), and black sea bass. This calendar was last revised in 2004, and Briggs reminded fishermen that they are responsible for checking the New York recreational fishing regulations (regarding size, number, and season) that currently apply. Those values tend to be changed every year (recreational saltwater fishing limits set by NYDEC https://www.dec.ny.gov/outdoor/7755.html).

New York State Department of Environmental Conservation

https://www.dec.ny.gov/outdoor/7755.html

New York State Department of Environmental Conservation establishes recreational and commercial fishing regulations.

Northeastern Fisheries Science Center
https://www.fisheries.noaa.gov/about/northeast-fisheries-science-center
> The Northeastern Fisheries Science Center's mission objectives include research and monitoring of the fishery resources of the Northeast United States, scientific advice, education, and outreach.

Odessey Expeditions
https://marinebiology.org/fish.htm
> This site offers a short course in ichthyology.

Reefs.com
https://reefs.com/author/toddgardner/
> Todd Gardner is an aquaculture professor and the author of articles describing his research and experiences collecting, keeping, and culturing marine organisms. Two examples are "The Serranids of New York" (2015) and "The Jacks of New York" (2015), in *Advanced Aquarist*, online at Reefs.com.

Squid Trawl Network
www.squidtrawlnetwork.com
> The Fisheries Team of the Cornell Cooperative Extension of Suffolk County maintains several websites in support of New York's fisheries. Information on these sites includes current and previous conservation gear technology projects; promotion of local seafood; bycatch avoidance; and various other information on New York's fisheries.

United States Fish and Wildlife Service, Northeast
https://www.fws.gov/northeast
> The Fish and Wildlife Service is the Department of Interior's agency emphasizing fisheries conservation.

Wikipedia
https://www.wikipedia.org/wiki/fish
> This site is a free encyclopedia of many aspects of fish biology.

Index